DEEP FREEZING

Marika Hanbury Tenison was born in London in 1938. She is a well-known authority on cooking and is cookery editor of the UK *Sunday Telegraph* and *Spectator*. She regularly contributes to other leading journals and makes frequent television and radio broadcasts.

She lives on a farm on Bodmin Moor in Cornwall with her husband, explorer Robin Hanbury Tenison. Their house and garden, including the kitchen, have been featured in *Vogue*, *Ideal Home*, *American House and Garden* and *Freezing Digest*.

Her other cookery books include *Deep Freeze Cookery* (Pan), *Deep Freeze Sense* (Pan), *Left Over for Tomorrow* (Penguin), *Best of British Cooking* (Hart-Davis), *Recipes from a Country Kitchen* (Hart-Davis) and *New Fish Cookery* (Hart-Davis).

TEACH YOURSELF BOOKS

DEEP FREEZING

Marika Hanbury Tenison

Illustrated by Vanessa Pancheri

TEACH YOURSELF BOOKS
Hodder and Stoughton

First Impression 1979
Second Impression 1982

Published in the USA by David McKay & Co Inc.,
750 Third Avenue, New York, NY 10017, USA
(British Library data)

British Library C.I.P.

Hanbury Tenison, Marika
Teach yourself deep freezing.
1. Home freezers
2. Food, frozen
I. Title
641.3 TX610

ISBN 0–340–22257–3

Printed and bound in Great Britain for Hodder and Stoughton Educational, a division of Hodder and Stoughton Limited, Mill Road, Dunton Green, Sevenoaks, Kent, by Richard Clay (The Chaucer Press) Ltd, Bungay, Suffolk

Contents

Conversion Tables

WEIGHTS

½ oz	10 g (gramme)
1	25
1½	40
2	50
2½	60
3	75
4	110
4½	125
5	150
6	175
7	200
8	225
9	250
10	275
12	350
1 lb	450
1½	700
2	900
3	1 kg 350 g

VOLUME

2 fl oz	55 ml
3 fl oz	75
5 fl oz (¼ pint)	150
½ pint	275
¾ pint	425
1 pint	570
1¾ pints	1 litre

(2 pint basin = 1 litre)

TEMPERATURES

Mark 1	275 F	140 C
2	300	150
3	325	170
4	350	180
5	375	190
6	400	200
7	425	220
8	450	230

MEASUREMENTS

⅛ in	3 mm (millimetre)
¼ in	½ cm (centimetre)
½	1
¾	2
1	2·5
1¼	3
1½	4
1¾	4·5
2	5
3	7·5
4	10
5	13
6	15
7	18
8	20
9	23
10	25·5
11	28
12	30

All these are *approximate* conversions, which have been either rounded up or down. In a few recipes it has been necessary to modify them very slightly.

You must either follow Metric or Imperial measurements. Do not mix the two in one recipe.

Temperature Chart

Celcius	Fahrenheit		
4° – 8°	40° – 48°		Temperature of domestic refrigerators
0°	32°		Freezing point of water
–5°	22°	*	Frozen food can be stored for 1 week
–12°	10°	**	Frozen food can be stored for up to 1 month
–17°	0°	***	Frozen food can be stored for up to 3 months
–21°	–6°	****	Freezing capacity for quick freezing of fresh food. Normal temperature
–25° to –30°	–15° to –22°		Temperature when quick freeze switch is on

NB: Temperature conversions are approximate.

Introduction

In our grandmother's day, women had almost all the time in the world to shop for food and to cook it. Now most of us hold down a job as well as being a wife and mother so time is always at a premium and a deep freeze saves time. With a deep freeze in your home you can buy food in bulk when you have the opportunity and then keep it in perfect condition in your freezer; you can cook double the quantity of one dish in the same time as it would have taken you to cook enough for one meal and freeze half to produce a good meal at a moment when you have neither the time nor inclination to cook a meal from scratch. By careful planning it is possible to prepare quite large quantities of food at one go so that when you want to entertain or enjoy yourself with your family, the time you have to spend in the kitchen will be considerably reduced.

All of us have to keep a careful check on our housekeeping budgets and here again the deep freeze can help. Buying meat, poultry, fruit and vegetables when they are at their cheapest and at the height of their season and then freezing them can produce a really considerable saving throughout the year. Freezing home grown produce, even if it is only a few humble bunches of parsley from a pot grown on a windowsill all helps to pare down living costs and sensible bulk buying can stretch a budget in a surprisingly successful way. Although the initial outlay in buying a deep freeze is quite a considerable one, it should be possible to pay for both the original cost of the machine and its running costs within a year of buying it from the savings you have made.

A freezer will also help you save money by enabling you to utilise things you might normally have thrown away. Onion and other vegetable peelings can be added to stocks to give flavour, colour and food value. Stale bread can be crumbed and frozen in small quantities, so can the grated hard ends of cheese. Where packaging is concerned, cream and yoghurt cartons and other containers can be used for storing frozen produce.

With a deep freeze you have a complete home cold storage unit at your fingertips, an extension larder that enables you to plan your life in advance, cater for any emergencies and provide you with top quality food at a reduced price.

PART ONE: BASICS

I

Buying a Freezer

Choosing the right deep freeze

New models of deep freezers are brought out every year. Gone are the days when, if you wanted a freezer, you had to buy a redundant ice cream chest. Now you can take your pick from hundreds of models in all shapes, sizes and designs from refrigerators that are combined with freezers, miniature freezers and large cupboard freezers – the problem is which to buy in order to fulfil your own and your family's requirements. Think about the following points before you buy.

A cold storage chest, cold store or conservator is not a deep freeze and cannot be used for freezing fresh food. It merely stores ready-frozen foods. To get good results from freezing at home you must have a machine which can be brought down to levels of below –21°C (–5°F). If the machine cannot reduce the temperature to this level, freezing will be slow and not entirely successful.

Refrigerators with a frozen food compartment will store food that has already been frozen but will not freeze fresh produce successfully.

To be efficient a deep freeze should be filled to at least two-thirds of its capacity nearly all the year round.

Try to work out in advance the kind of freezer space you need. Obviously someone living in the country with a large vegetable garden and easy access to cheaper meat, poultry and perhaps fish will need a larger freezer than a small family living in a city where the shops are open all the week and shopping can be done day by day.

It is only possible to freeze fresh produce in 10% of the storage space at any one time, so if you plan to freeze meat in bulk, by buying a whole lamb or a quarter of beef, you can see that you will need quite a large storage space.

The amount of entertaining you do and the number of mouths you have to feed each day also have a bearing on the size of freezer you plan to buy.

The size range of freezers is now vast. They start at about 4 cubic feet (11 cubic m) and go up to over 20 cubic feet (57 cubic m) for domestic use. Manufacturers usually recommend a storage space of about 4 cubic feet (11 cubic m) per person. If you live in the country this would seem to me to be very much on the conservative side but it is probably a fair enough guideline for a family of about four living in a city.

Look at your friends' freezers and try to imagine how much space you would use yourself before looking at empty freezers in a shop; empty freezers always look enormous and it is often difficult to imagine how you could possibly fill all that gleaming space.

Siting your freezer

Obviously your first criterion will be to choose a site where you can afford the space for a machine as bulky as a freezer but do remember that the colder the area the less your freezer will cost you to run. Most freezer motors are even more quiet than a refrigerator when they are running so that noise is not really a problem. Plan to house your freezer as far away from central heating pipes, solid fuel burners or any cooker as possible and ask to have it fitted with rollers (in the case of a chest freezer it is worth buying an inexpensive low platform equipped with rollers for the freezer to sit on).

Choose a site that is near to a 13 amp socket so that the lead can be plugged in without difficulty.

If you are planning to make the fullest possible use of your freezer and not just to use it as an extension larder for the storage of fish, meat and vegetables, do not site it in a position too far from the kitchen. You will also need to have the site well lit so that packages can easily be identified.

The site must be level and if your freezer is to be placed in a garage or outhouse it is worth putting it on blocks to allow the free circulation of air around the machine.

If you decide to keep your freezer outside, in a garage or perhaps a tool shed, remember that you will have to walk down to the bottom of the garden in mid winter each time you want to take out a packet of peas. Keeping your freezer outside will cut down on the running costs but it can also present problems when, for instance, damp gets into the wiring and causes a short. Outside freezers should be checked every other day in wet weather and, if you live in a built-up area, it may also be necessary to ensure they are kept locked at all times.

The choice of freezers available

Star rating

Most freezers are sold with star symbols on the outside. These symbols are used by all manufacturers and show what temperature the freezer can maintain.

Note: The lower the temperature of your freezer the longer food can be stored in it. Black stars represent the length of time food can be stored in a deep freeze. A white star indicates that the temperature of the freezer can be lowered for freezing fresh food quickly to obtain minimum temperatures in the shortest possible time.

Freezer symbols

★	Frozen food can be stored for up to 1 week.
★★	Frozen food can be stored for up to 1 month.
★★★	Frozen food can be stored for up to 3 months.
★★★★	The freezer has a compartment for the fast freezing of food and food can be kept frozen for the length of time as indicated for various items: see chart on pages 34–43.

Upright freezers

Undoubtedly these are the freezers which look smartest, make finding packages of frozen food easier and are easier to fit into the modern kitchen which tends to be small and streamlined. Upright freezers usually have a special quick freezing compartment which can be boosted to lower the temperature for freezing fresh produce.

The cabinets are divided into shelves usually equipped with baskets for sliding frozen food in and out of the cabinet. On some models the inside of the doors also provides storage space which is ideal for small items.

Cabinet freezers come in all shapes and sizes. You can buy a small freezer cabinet to fit under working space in the kitchen, a large cupboard variety or one that is combined with a refrigerator (either on top of or alongside it). Although upright freezers do make the checking of deep freeze packages much easier they have four disadvantages:

1. They are more expensive to run than a chest or 'coffin' shaped freezer because the temperature of the freezer is raised each time the door is opened and warm air is sucked into the cabinet.
2. Space in the shelves is usually confined to about 12 inches (30·5 cm) height so it is not possible to store large and sometimes awkward pieces of meat.
3. The dividing shelves and the sliding baskets take up quite a considerable amount of room.
4. The shelves encourage icing up and upright freezers should be defrosted at least once every six months. (It is possible to buy automatically defrosting machines and I think this is something that is going to become more and more popular although, at the moment, automatic defrosting tends to add considerably to the cost of the machine.)

The chest freezer

These are boxes or coffin-shaped receptacles with a top-opening lid. For their size they are usually larger in storage capacity than an upright freezer and because warm air does not rush into them every time the lid is opened, they are cheaper to run. Chest freezers tend to be less expensive than the more sophisticated upright models and are more convenient for storing large, awkward packages in.

Chest freezers can be bought with a self defrosting element and they should have an area for fast freezing of fresh food.

The drawbacks of a chest freezer include:

1. It is necessary to dig down to the bottom to find packages required.
2. It is difficult to divide into compartments.
3. It is an awkward shape to site.

Running costs

Providing your freezer is situated in a reasonably cool site, the door is not opened more than is absolutely necessary (remember that opening the door or lid will probably make the motor cut in at once), and the freezer is kept to at least three-quarters full, it should not cost you more to run than a refrigerator of the equivalent size. Small freezers will usually use somewhere in the region of 2 units of electricity per cubic foot (28 cubic cm) a week but larger freezers will tend to use slightly less electricity per cubic foot (30·5 cubic cm) on average.

Remember that you are paying the electricity bill whether your freezer is full or empty so try to keep a steady flow of produce going through it all the year round to get the maximum return for your outlay.

The fast freeze switch on your machine boosts the motor and uses additional electricity so switch it off as soon as the produce you are freezing is solid all through.

A layer of ice on the inside of the freezer will cut down its efficiency and increase the running costs, so ensure the freezer is defrosted as often as necessary.

Shop around for your freezer

Points to consider:

1. Your freezer should have a special compartment for quick freezing of fresh produce.
2. Self defrosting mechanisms can save you a lot of time.
3. A light in the machine is a great advantage especially in the chest variety.
4. Has the machine got a warning light, buzzer or alarm? This will show you when the temperature has risen to a dangerous level and is, to my mind, an essential piece of equipment. Small alarm systems, working on a battery, can be added to your machine for a relatively small cost.
5. Watch out for shop-soiled bargains. Two of my freezers over the years have been models which have been displayed at the local County show and then sold off as 'slightly soiled' or 'demonstration' models. This can bring down their prices very considerably and there never seems to be any sign of the 'slight soiling' whatsoever. This kind of reduction in price can also apply to models which have been sitting in a shop window for some time, which get a slight scratching on the surface or which have got dirty in storage. Keep an eye out for bargains of this sort.
6. Watch out for 'special offers'. These are often because a model has been replaced by a more advanced machine but since this 'advance' usually only applies to the look of the machine and not to the all-vital working parts it is a matter of aesthetics rather than practicability. Good bargains are usually to be found in sales but check that your freezer is made by a

reputable manufacturer and that spares are available for such things as motor parts and the rubber sealing that goes round the doors or lids.

7. You will usually find you get a good discount for cash but make sure the deal includes a guarantee for the first year and that insurance and maintenance are available. I have never had a freezer go wrong, but I have heard of others that have and it is important to get good service within 24 hours if that happens. If you buy on hire purchase try and make a deal that includes both maintenance and insurance.
8. *Secondhand freezers:* There is always a big turnover in secondhand deep freezers because so many people discover after a fairly short time that they need a larger capacity than they had at first decided on. Secondhand freezers are usually quite reasonable in price but make sure you buy one from a reputable dealer and can get a service agreement on it.

If you wish to buy a larger freezer yourself you can usually get a fair trade-in price for your old one and this is usually a better deal than trying to sell it on your own account.

2

Maintaining your Freezer

The new machine

Before you begin to fill your freezer wash it out with hot water to which a little vinegar or bicarbonate of soda has been added and dry it carefully with a soft cloth. Switch on; set the dial to its lowest setting or switch on the boost switch and let the machine run for at least twelve hours before stocking it. If your machine has a top at working level it is well worth-while protecting the surface by covering it with a stick-on vinyl material; even if the machine is in a garage you are bound to want to put the odd thing on top of it every now and then and a scratched surface will lower the value should you want to trade-in the freezer for a larger model.

Test the alarm system to make sure it is working and make a point of always testing it at least once a week.

Defrosting

However careful and efficient you are, packages do get lost in your freezer and some get forgotten about for longer than their freezer life, so make a point of defrosting your machine at least once every six months to ensure that goods are rotated, that storage safety dates are adhered to and that pierced or damaged packages are given first aid.

Even if your freezer has a self defrosting unit it is well to clear out the freezer once every six months in order to keep a comprehensive track of the contents.

Defrosting procedure

Allow yourself plenty of time and wear clean gardening gloves for this chore as it gets very cold deep inside a deep freeze. Place a blanket on the ground near the freezer and spread newspapers on top of the blanket, or have ready a supply of insulated paper sacks which most deep freeze supermarkets supply with your purchases –although should it take too long to defrost your machine some more delicate items might be damaged by even the slightest thawing.

1. Switch off the machine and disconnect the plug before starting work.
2. Take out all the frozen packages and place them on the newspaper on the rug or in the insulated paper bags. Set all damaged or outdated packets on one side. Fold the edges of the blanket over the frozen packages.
3. Place a plastic bowl of very hot (not boiling) water inside the freezer and begin to knock off, or slide off, the frozen ice from the sides of the machine. Do not use a sharp knife or instrument as this will damage the deep freeze. I find a wooden spatula ideal for the purpose of scraping or lightly knocking off the ice.
4. Use a clean brush and a dustpan to sweep the ice that falls from the sides. Use a large sponge to mop up any water collecting on shelves or on the bottom of a chest freezer.
5. Wipe out the frost free interior of the freezer with a cloth dipped in warm water to which a tablespoon of bicarbonate of soda or vinegar has been added to every quart of water and then wring out and finally wipe over again with a soft dry cloth.
6. Switch on your machine to maximum and switch on the fast freezing switch if you have one. Leave to run for thirty minutes and then carefully replace all the packages making sure they are divided into sections and that earlier dated packages are nearest to hand.
7. Repack any broken packages and, if they have suffered from *freezer burn* (extensive build up of ice crystals inside the packaging or white patches appearing on the surface of frozen produce), make a point of cooking or using the contents as quickly as possible. If it is not possible to use them at once repack the packages in fresh packaging material wiping off any ice particles before packing.

3

Planned Loading of your Freezer

When you are pressed for time and produce from the garden has to be frozen in a hurry while it is still fresh, or when you return home from a bulk buying visit to a freezer centre, you may have a tendency to put things into the freezer in no particular order. Do not give way to this temptation because a well organised and well packed freezer will save you considerable frustration in the long run. Divide a chest type freezer into compartments by means of wire dividers which can be bought from a deep freezer equipment manufacturer and keep fruit, vegetables, meat, poultry, made up dishes etc., separated. Keep smaller items together so that they do not get lost amongst larger packages and note down the position of items in your freezer record book. Upright freezers with shelf space are easier to keep in order as each shelf can be used specifically for one group of produce.

Try to keep packages as uniform and box-shaped as possible in order to save space.

Loading chest type freezers

Use the bottom of the freezer for bulky, unwieldly packages and for items which are destined for long term freezing. If you have large quantities of home grown vegetables, items you have bought in bulk or a number of chickens etc., it is worth packing them together in a plastic carrier bag or large polythene bag.

Items which are to be used quickly should be kept near to the top of your freezer together with items which may be easily damaged even though they are frozen (for example, cakes and soft fruit which can be damaged if heavier objects are put on top of them). Wire baskets are usually easily obtainable for packaging produce which is to be kept at the top of your chest freezer. The more you can manage to divide your freezer into compartments the easier produce will be to find in a hurry.

Loading cabinet freezers

Pull-out shelves make this an easier job than loading the chest variety of freezer but again the accent should always be on dividing your shelves into produce of the same variety and, if necessary, overpacking groups of the same produce in plastic bags. If the door of your freezer has storage space keep this for small items like cubes of concentrated stock, small packages of herbs, grated cheese, breadcrumbs etc.

Pack your produce in as neat and uniform a package as possible and rotate the produce in your freezer regularly so that packages from the back of the shelves which may be reaching their deadline are brought out towards the front of the shelves.

Useful headings under which to divide produce in your freezer

Meat; offal; poultry; fruit; vegetables; 'emergency foods' (i.e. commercial products which are quickly cooked such as fish fingers, hamburgers etc.); home-made pre-cooked dishes; cakes, bread and other baked products; soup and sauces; small packages for flavouring (i.e. herbs, home-made tomato purée, citrus fruit peel etc.); produce to be used within a week.

Remove all produce from the fast freeze area as soon as it is frozen so that you always have the area clear for more freezing of fresh produce.

General notes on the quick freezing of fresh and pre-cooked foods

The faster food is frozen the better the results will be when the food is removed from the freezer. Almost all modern freezers have a special compartment, next to the motor, for the fast freezing of food and a 'fast freeze' button which lowers the temperature of the freezer for as long as it is necessary to freeze the food until solid. Switch this button on for about two hours before freezing food and remember to switch it off as soon as the food is frozen. The time it will take for food to become solid depends on how much food is being frozen at one time and on the temperature of your individual freezer.

Do not attempt to freeze too much food at one time as this will raise the temperature in the freezer and it will take longer to freeze the produce until it is solid. If you are freezing a whole lamb, for instance, or a quarter of beef, freeze the meat in shifts of relatively small quantities and wait until each batch is frozen solid before putting more unfrozen meat into the freezer.

4
Buying in Bulk

Buying in bulk is often difficult for those who have no experience in this field and mistakes can be expensive. Like most things, though, buying successfully in bulk just needs a little careful thought and planning and if it is well done it can lead to a saving in both time and money. Buying in bulk for the freezer is merely an extension of buying in bulk from a cash and carry for the larder.

Four points to check before you buy anything in bulk

1. Are you buying the produce just because it is on offer or a 'cheap' buy or is this something you and your family really want?
2. Are you sure the quality of the produce is good or is it reduced because the manufacturers or producers haven't been able to sell it any other way?
3. Does the space in your freezer justify a bulk purchase of this sort, i.e. will the purchase of a large quantity of meat at one time mean that you will not be able to freeze more valuable products such as vegetables and fruit?
4. You may have found items such as beefburgers and cod fingers good value in the past but will you really want to enjoy them frequently during the next month?

Advice for beginners

There is always a temptation to rush out and buy enough food to fill your freezer as soon as you have it installed. Don't. Take things slowly at first, shop around and buy small quantities of frozen food before you decide which are worth buying in bulk.

Choosing your source for bulk buying
Buying frozen food is no different from buying any other produce. You have to shop around for it to find good quality at the cheapest price.

Most large towns now have deep freeze supermarkets and these often have cheap offers for bulk buys but buying from them may not always be the least expensive way of getting the goods you want or value for your money. For example, cheap hamburgers may be made of more rusk than meat.

Cheap offers are usually of goods that sell under a brand name you do not recognise and there may be a very good reason why they are being sold off in a hurry so be very wary of buying large quantities of unknown produce cheaply. On the other hand most of the big frozen food supermarket chains carry their own brand name on a proportion of their goods and these usually are reliable and good value. Bright colourful packaging does not necessarily indicate the quality of the goods inside and that, in the long run, is what counts. If you have more than one deep frozen food centre in your area, take the trouble to compare prices and quality.

Cleanliness and good packaging is of paramount importance in all forms of freezing and it is worth taking a good look around your local deep frozen food centre to check that the freezers are kept immaculately clean, all damaged packages are removed immediately and that stock is being rotated all the time. Never buy anything that is being sold in a damaged wrapping and watch out carefully for signs of freezer burn or excessive crystallisation in transparent packages. Remember too that the supermarket owners want *your* custom. If you do not feel that goods are up to standard do not be shy about complaining and always ask for advice from one of the attendants if you feel you need it (many of the chains have home economists on tap for just this purpose).

Buying fresh food in bulk

If you take advantage of seasonal gluts and abundance it is possible to make considerable savings by buying fresh products in bulk and freezing it yourself. Remember, however, that the temperature in your freezer is not as low as that of the temperatures used in commercial freezing, so your produce will not have such a long life as that of a commercial product, nor will it, in some cases, be of such good quality.

The produce you freeze can never improve by being frozen so do not be tempted to buy inferior quality produce just because it is cheap.

Think twice before buying fresh produce in large quantities. Can you really cope with the freezing of a whole sheep or lamb yourself or would it, perhaps, be better to spend a little more and have half a lamb now and another half in six months' time? Brussels sprouts may be in plentiful supply this year and cheap for a green vegetable but do you really want to be eating Brussels sprouts throughout the summer just because they happen to be cheap in November?

Good packaging is almost as important as the quality of the produce itself so make sure you have the equipment and the time to be able to process fresh food for the freezer before you buy it in large quantities and remember also that the fast freeze section of the freezer can only cope with so much fresh food at one time.

Buying fresh meat in bulk

One of the most important factors here is to think about the butchering of the meat. Butchering is a highly specialised business and not something to be undertaken lightly in the home. I once bought a lamb, skinned and cleaned but otherwise in one piece, without stopping to think what I was actually going to do with it when I got it home. It took me two days working practically full time to saw and cut the carcass into pieces and when I had finished it was almost impossible to tell a leg from a piece of the best end. I learned then that it was not worthwhile trying to short circuit the system as I had neither the equipment nor the knowledge to butcher the animal successfully. Good butchering leads, in the long run, to economy and it certainly makes life a lot easier for both the cook and the carver.

Do not buy more meat at one time than you can cope with. Most freezers can freeze down about 10% of the capacity in a 24 hour period. This means that any meat which cannot be accommodated immediately will have to be kept chilled or at least put in a cool place until it can be frozen.

An average carcass consists of about 30% wastage in fat and bones etc. so take that into your calculations. If you have access to a source where you can buy whole carcasses in bulk find a friend or neighbour who is prepared to share the bargain with you and check with a butcher before you buy the meat to find out what he will charge for butchering the animal.

Try to freeze the meat you buy in quantities that are going to be suitable for the size of your family and your way of life. Barons of beef and saddles of lamb sound attractive but are you really going to have a suitable occasion when that quantity of meat will be eaten at one sitting?

Beef and lamb should be hung before being frozen so make sure this has been done. Meat that has not been hung will be lacking in flavour and tend to be tough.

Keep your freezer stocked with beef, lamb and pork rather than just having one kind of meat which can lead to a rather monotonous diet.

Make sure you divide your chops, steaks or other small cuts with foil, freezer paper or transparent wrapping film before packaging it and freeze in small rather than large quantities.

It is advisable to mince meat and to cut it into cubes for stewing before freezing rather than to have to wait for it to thaw out before processing it. Ask your butcher if he is prepared to mince the meat for you as this is a very arduous job to do at home by hand. He may also agree to make sausages for you if you are buying pork in bulk (a good idea as you can then alter the seasoning and flavouring of the sausages to suit your taste).

Buying fresh fish in bulk

Fishmongers will talk about fish being 'fresh' but this often means it has been caught far out at sea, frozen as soon as it has reached the fishing boat and then thawed out before being put on sale. Fish that has been frozen on board and then thawed out will not necessarily be unsafe to eat if it is frozen again but it will probably lose some of its taste and texture.

If you want to freeze a quantity of fresh fish at home, it is best to buy it from a fish auction or fish market, but remember you must freeze the fish as quickly as possible and you may also have to gut the fish before freezing.

For the best results, fish for freezing should be as fresh as possible and never damaged or bruised in any way.

Small fish or fillets of fish should either be individually wrapped in polythene film before being overpacked or the fish or fillets should be spread on a flat tray and 'free' or 'open' frozen until solid before being packed. This will help to keep the fish or fillets separated and prevent solidifying into a block.

Fish is extremely susceptible to bacteria so make sure your working conditions and your hands are scrupulously clean before beginning to prepare it for the freezer.

As fish is a seasonable commodity take the trouble to check and see whether a particular variety really is in season before buying it in bulk. It also has a shorter life in a home freezer than meat so consider how much your family actually do like eating fish and fish dishes before buying it in large quantities. Having mackerel everyday on a seaside holiday is one thing but eating it everyday at home may not be quite such fun.

Shellfish can be dangerous unless absolutely fresh. Only buy shellfish from a reputable source; make sure it has not been frozen and then thawed out, and freeze it as quickly as possible when you get it home.

Buying fresh poultry in bulk

Some food supermarkets sometimes have good value for money bargains in birds that are 'fresh not frozen'. The low price does not reflect poor quality but is due to seasonal conditions. These birds make good buys. Freeze the birds as soon as possible and take great care to ensure that they are really well packaged and that as much air as possible has been forced out of the bags. The giblets should be frozen separately unless the birds are only to be stored for a short length of time.

Most chicken pieces on sale have already been frozen and thawed out and should not be re-frozen. If you like keeping a store of chicken pieces in your freezer make sure you buy them from a reasonable source or buy whole chickens when the price is right and joint them yourself using the carcasses to make stock.

'Boiling' fowls. These are often laying birds which are past their prime and they can make inexpensive and well flavoured dishes, casseroles and pies.

When buying 'undressed' birds (poultry which has not been prepared for the oven) look out for smooth, pliable legs with fine rather than coarse scales, a pliable backbone and breastbone, good plump legs and breast with the skin over the breast a good white colour rather than yellow. Fresh birds should have clear eyes and do not buy any that have an unpleasant or slightly fishy smell.

Buying fresh vegetables in bulk

If you plan to buy fresh vegetables in bulk for the freezer make sure they are in prime condition when you buy them, prepare them for freezing as quickly as possible and freeze them in as short a time as you can after buying them. Most vegetables should be 'free frozen' or 'open frozen' before being packed and you will only be able to cope with so many at a time in the fast freeze section of your freezer. Good buys can be made when vegetables are in season at a 'pick yourself' farm but remember the vegetables will have to be processed before they can be frozen and there is no point in buying more than you can cope with at one time in your kitchen.

Cost out the price of vegetables bought in bulk carefully before buying them. The wastage factor is often quite high and you may find that, in the long run, you would in fact have done better to have bought the same vegetable ready frozen in bulk from a deep freeze supermarket.

Good buys can often be made from your local greengrocer on a Saturday evening. He may well sell you a tray or box of vegetables at an astonishingly reasonable price if he has overbought and risks having the vegetables go off over the weekend. Bruised or overripe tomatoes can sometimes be bought for very little and are invaluable as a source of home made tomato purée.

Try to get a good variety of home frozen vegetables rather than concentrating on freezing too many of one kind.

Pack vegetables in small rather than large quantities. You can always take out more than one package but opening a large package for a small amount of vegetables is bad freezing practice.

Buying fresh fruit in bulk

Select only fruit that is in prime condition and if it has been grown in the UK try to ensure that you freeze it within 24 hours of it having been picked. Again, do not buy more fruit than you can cope with at one time. A tray of peaches, for instance, may seem like a very good bargain but remember that you will need the time to skin them and pack them in a sugar syrup to prevent discolouring.

Soft fruit and berries should all be 'free' or 'open' frozen before being packed so make sure you have shallow trays for this purpose before investing in large quantities of soft fruit. Also, as it damages easily you will need rigid containers to pack it in.

'Pick yourself' fruit and vegetable farms sometimes offer remarkably good value in fruit sales particularly if it has been a 'glut' year but remember that, ideally, fruit for freezing should be picked first thing in the morning after the dew has dried but before the sun has gained its full strength. Check prices with those in the shops and markets before you break your back picking the fruit yourself and do not leave picked fruit in the sun for a second longer than you need to. Process and freeze the fruit as quickly as possible after you have picked it and try to grade the fruit as you pick so that it will all freeze at the same rate.

Jam is expensive and you can often buy second grade or over-ripe fruit for a reasonable price with which you can make your own home made jam. The fruit can be prepared and frozen and the jam made at a later date.

No fruit will improve by being frozen so make sure the produce you freeze is fully ripened and free from blemishes or briuses. Do not freeze fruit that has been picked when it is damp or rain sodden.

Other fresh food commodities that are worth while buying in bulk

Bread: For convenience bread can be bought in bulk and successfully frozen but it does need to be well packed as all breads, pastry and cake products are very susceptible to 'freezer burn'. Wax paper wrapping on sliced bread is not adequate packaging material and the loaves should be overpacked in polythene bags. However, remember bread is a bulky commodity so do not lose valuable freezer space unless there is a good reason for doing so.

Butter: If your family are hearty butter eaters you can make a good saving by buying butter in bulk from a 'cash and carry' store and freezing it for up to six months.

Eggs: It is surprising how wide a price range there is in the cost of eggs over a twelve month period and at times their price can fall very sharply indeed. Watch the egg prices in your newspaper and buy up a stock when they are at their cheapest to freeze for cooking. The eggs should be lightly beaten with a fork before freezing and either a little seasoning or sugar should be added to the eggs to provide a stabiliser.

Catering size tins
Buying catering size tins can be a great saving providing you can store the residue of the contents. Tinned fruit, vegetables and soup etc. can all be frozen successfully.

Buying frozen food in bulk

The quality of commercially frozen foods from vegetables to ready to serve or cook dishes has improved enormously in the past few years but, like all convenience foods, they can be on the expensive side and some brands are of a very much higher quality than others with prices not always proving to be a guideline as to quality. Buying in bulk from one of the deep freeze centres can help to cut down on the cost providing you observe a few guidelines.

Bulk packs of frozen produce often come in quite large quantities so try and test a product first by buying a small quantity to see if you and your family really like it before getting a large quantity at one time.

Beware of cheap offers
Of course it is worth looking out for bargains whenever you go shopping and there may often be a very good reason why the goods have been reduced. Cheap offers may be a genuine bargain to attract customers and get rid of a glut of stock but they could, on the other hand, be produce that has been sitting around for a considerable time or that may be of an inferior quality and therefore has proved difficult to move from the supermarket shelves. A bargain is no bargain at all if you get it home only to find it taking up valuable room in your freezer. Remember that by buying in bulk you are, in effect, tying up capital so make sure you get your money's worth.

Packaging must be airtight
If you buy fishfingers or hamburgers etc. in bulk packs their quality will deteriorate once the pack has been opened. As soon as you get the produce home, divide it into family sized portions and wrap the portions separately, carefully labelling the amount the packages contain. If you open a large package of fruit or vegetables

and only use part of the contents, make sure you seal the package carefully before returning it to the freezer.

Thawing and re-freezing causes deterioration
Get your bulk purchases home as soon as you can after buying them and put them into your freezer the moment you get back to the house. Most deep freeze supermarkets now provide insulated paper sacks for this purpose and these can be kept and reused. If you live a long way from a freezer centre, you may find it worth while investing in a large insulated box which will keep your food frozen during the journey home. When you are packing frozen food into one of these containers fill up any airspace in the container with crumpled newspaper.

Keep your cool when buying in bulk
Deep freeze supermarkets can be like an Aladdin's cave of temptation to even the most level headed of housewives. You go into them meaning to buy a pack of beans and perhaps a carton of fish fingers for the children and you suddenly find yourself filled with a desire for middle cuts of salmon, a bargain in steak, some chocolate éclairs that you think just might be invaluable and a whole lot of other mouth watering looking packs that you really do not need at all. It is very easy to get carried away with the wide choice of delicious looking foods you see on offer especially if you are new to the game of freezing. Write out a list of those goods you really intend to buy before you go shopping and stick to a realistic budget at all times.

Check the dates on commercially frozen produce
More and more produce these days is labelled with a 'selling' date or a date by which the product should be used. These tend to be guidelines only and do not mean that the produce is not fit to eat once the date has expired. Nevertheless they are there for a purpose and you may well find different dates on packages in the same freezer unit. In well run supermarkets the latest dates are at the back of the cabinet so if you plan to keep a certain product for some time check the dates on the packages.

Check the ingredients of the contents in a package before buying it
Ready to cook or serve dishes should have the ingredients clearly listed on the outside of the packages and it is wise to check these.

Buying frozen meat in bulk

Check the prices carefully as I said before. Supermarket chains now often have offers of sides, quarters, legs or other cuts of meat on sale in bulk and these can be good buys. But check the price against that of the same meat in your butcher before laying out a considerable amount of money and also examine the packages carefully to see that the meat has not been affected by freezer burn and that it is well wrapped and properly packaged.

The advantage of buying your meat ready frozen in bulk rather than buying it fresh and freezing it yourself at home, is that factory blast-freezing processes freeze down to a far lower temperature than can be achieved in a home freezer. They freeze at a very fast rate preventing a large amount of tissue breakdown and providing an end product that is as near as possible in quality to that of a first rate fresh product. However, do not buy more meat than your family are going to be able to enjoy just because it seems like a good buy.

Check through the packages and make sure that chops and steaks are in 'free-flow' packs and that minced or diced meat is not packed in larger quantities than you can cope with. Check also that joints are of a size you usually cook.

Buying frozen fish in bulk

Commercially caught fish is frozen within hours of catching and, providing you are dealing with a reputable firm, the fish you buy in the frozen state should be fresher than any you actually buy, take home and freeze yourself. There are, however, various points that should be taken into consideration when you are buying a quantity of frozen fish to store in your freezer.

Do not buy too much fish at one time. Frozen fish has a fairly short freezer life if it is to be served when it is in prime condition and it is best to avoid buying more than a two or three month's supply of fish at a time.

There is a considerable traffic in the sale of below quality made up fish products such as fish fingers and fish cakes etc. The quantity of fish used in these products should always be clearly marked on the outside of the package and you should check on this and then select the product with the highest proportion of fish in the contents of the made up article.

Fish is one of the most sensitive products that can be frozen and therefore needs even more careful packaging and sealing than almost any other fresh produce. Check packages over carefully to ensure they are undamaged before buying them and do not buy packs of fillets or steaks that have not been frozen individually or in a 'free-flow' pack; once portions of fish have been frozen they are impossible to separate until they have been completely thawed.

Because of overfishing and other supply problems, 'traditional' fish are becoming scarcer and many new varieties are now being tried. Many of these varieties are not only good value for their price but also excellent in both flavour and texture providing a wealth of material on which to base some really exciting dishes. Do not ignore a species of fish just because you may not recognise its name and keep your eye open for offers of such varieties as coley, huss, monkfish, sprats, squid and, of course, mackerel which is quickly taking the place of the now hard to get herring as an inexpensive, everyday fish.

Buying frozen poultry in bulk

Buying frozen poultry in bulk can present quite a considerable saving if your family prefer chicken to meat and since there are many firms who specialise in rearing and processing birds for the freezer they can often work out far less expensive than fresh birds. This particularly applies to poultry for Christmas.

All poultry has a large percentage of bone in comparison to meat so take this into consideration when you are comparing the price of poultry against that of meat. The larger the bird the more economical a bargain it will turn out to be. Look out for bargains in chicken portions as these can often turn out to be very reasonable indeed. I find chicken wings a particularly good buy as they do not seem to be all that popular, are usually relatively inexpensive and are invaluable for making a number of exciting dishes.

If you are buying a whole bird frozen make sure you are not being cheated out of the giblets which are all important for the making of stock or gravy. Some firms sell these separately without lowering the price of the bird to compensate for the loss.

All poultry is particularly susceptible to freezer burn so check packages carefully to ensure they have not been damaged or pierced in any way and avoid any birds that have the tell-tale white patches which denote freezer burn.

Buying frozen vegetables in bulk

If you do not grow your own vegetables and have to rely on your greengrocer and on buying vegetables when they are at the height of their season and therefore at a reasonable price, then it makes good sense to supplement your fresh vegetables with some bulk buying of frozen produce. Freezing has practically no effect on the nutritive value of vegetables and therefore, providing you are buying produce that has been frozen at the right time, carefully prepared and packed, you are probably getting a better deal than you would if you bought slightly inferior produce in a market or from a greengrocer and then had to peel and prepare it yourself.

Buy only those vegetables that your family really do enjoy and do not be tempted to buy too much at a time. Inevitably the larger the pack the cheaper the price of the vegetables it contains, but remember that large, bulky packs take up valuable storage room and once they have been opened their life is shortened. If you do buy very large packs it is worth dividing the packs into smaller quantities as soon as you get them home but this must be done quickly so that the produce does not have time to defrost before it is repacked.

Compare the price of frozen vegetables with that of fresh if they are in season and do not fall into the trap of eating frozen vegetables that have actually cost you more than the same produce being sold locally fresh from the soil. Check packages minutely to ensure they are undamaged and refuse to buy any packages that are pierced or broken in any way. Be wary about very cheap buys as they may be produce that has been reduced in price because its quality is below standard.

Buying frozen fruit in bulk

A large number of fruits freeze exceptionally well and as they are grown, packed and frozen on such a large scale frozen fruit can often work out to be less expensive than the fresh product. Good as it is, however, frozen fruit has not got that indefinable perfection of flavour that is to be found in something freshly picked and still warm from the heat of the sun.

Make sure all fruit has been 'free-frozen' and is not packed in a solid lump and do not disregard packs that carry a name you have not heard of as these are often less expensive than those packs bearing a well known name and they need not necessarily be of a lower standard. When the fruit season has been good and there are gluts of soft fruit bringing down the cost it can make sense to make your own jams from frozen fruit.

5

What is most Suitable for Freezing?

Your freezer is not a machine of magic; some fresh produce and cooked foods do not freeze as successfully as other items so concentrate on those commodities which you know are going to be a success rather than risking disappointments at a later date.

These are the 'Super Freezers' that have a considerable freezer life.

Meat: All lean joints including beef, mutton, lamb and pork; chops and steaks; minced meat.

Fish: All whole fish and fillets or cutlets of white and river fish freeze well.

Poultry and game: All poultry and game freezes well, either whole or in joints, providing the giblets which have a shorter freezing life are stored separately from the birds.

Vegetables: With the exception of salad ingredients such as lettuce, radishes, spring onions (unless they are chopped), unblanched chicory etc., vegetables freeze very well indeed, best freezers of all are beans, peas, spinach, broccoli, asparagus and Brussels sprouts.

Fruit: Raspberries, gooseberries, currants, rhubarb and blackberries can all be relied upon to freeze successfully and to emerge from the freezer with almost exactly the same taste and texture as that of the fresh produce. Melon and pineapple are good long storage freezer material provided they are packed in sugar or a sugar syrup.

Pastry: Supplies of shortcrust, flaky and puff pastry keep well in the freezer so make double or treble the quantity you need when making a pie and freeze the rest in usable quantities. Most commercial brands of frozen pastry are extremely good indeed and well worth keeping a stock of for emergencies or convenience.

Other foods that freeze well but have a less extended freezer life

Meat: Composite meats, sausages, bacon, ham and offal.

Fish: Smoked and oily fish, whole or in fillets, and shellfish.

Pre-cooked foods that freeze well

Commercially pre-cooked and ready frozen foods.

Bread, cakes and cookies: Rolls, buns and cut or uncut loaves, both brown and white, are all excellent freezers as are large or small cakes. Uncooked biscuit mixes can be shaped into a roll, frozen and then thinly sliced and baked from the frozen state providing quickly cooked cookies at short notice.

Meat dishes: Stews, casseroles, meat loaves, meat sauces, vegetables and pancakes stuffed with a meat filling, pâtés, terrines, meatballs with or without sauce, hamburgers and meat patties, meat puddings and pies. Leftover roast or cooked, sliced meat can be frozen provided it is covered with gravy or a sauce of some kind.

Poultry: Poultry casseroles, stews and pies all freeze well and slices of cooked poultry can be frozen provided they are covered with a sauce of some kind.

Fish cakes, pies and soups: Fish forcemeat to use as the base of fish soufflés, fish mousses, fillets of breadcrumbed or battered fish that have been partially cooked and can be deep fried from the frozen state.

Vegetable dishes: With the cost of meat rising more and more housewives are having to supplement their menus with vegetable dishes and there are a wide variety of exçiting vegetable main course dishes which adapt well to being frozen.

Pastry: Filled or unfilled, baked or unbaked sweet and savoury flan cases can all be frozen with good results.

Batters: Pancakes of all kinds, filled or unfilled, all freeze well.

Stocks and soups: A good supply of stocks of varying kinds are invaluable if you are to get the best out of your freezer; use all carcasses and leftovers to provide meat, poultry and fish stocks for instant use.

Most soups freeze well and can be made in bulk and frozen for later use.

Puddings: There are a wide variety of pre-cooked or prepared puddings which you can make in advance and freeze successfully including ice creams, many fruit based sweets, sweet mousses and iced soufflés, fruit flans, tarts and pies, 'Alaskas' ready for cooking and gâteaux.

Produce that does not freeze well

Hard boiled eggs: Whole or chopped hard boiled eggs go rubbery if they are frozen. The only method of freezing hard boiled eggs is to mash or purée them and incorporate them into a cooked dish.

Whole fresh eggs: Eggs should be separated, the yolks should be lightly beaten until smooth before being frozen and both the whites and yolks benefit from the addition of a little salt or sugar added to them as a stabiliser.

Cooked old potatoes: Do not attempt to freeze dishes such as a Lancashire Hot Pot which include chunks of cooked old potatoes as these tend to harden and go rubbery in the freezer.

Overcooked dishes: Dishes that are already overcooked should not be frozen as they will have to be thawed and re-heated after being removed from the freezer and their flavour will deteriorate all the time. Freezing produce can *never* improve its taste and texture. Pre-cooked dishes that are to be frozen should be undercooked.

Overspiced dishes: Some spices and herbs increase in strength during their freezer life; avoid this by using a light hand with herbs and spices when cooking for your deep freeze – you can always add extra flavouring when the dish is re-heated.

Sauces and soups etc. that are made with cream: Soups and sauces that are inclined to separate in the normal course of cooking will certainly do so in the freezer. If you are making soups or sauces with egg yolks and/or cream it is better to add the egg yolks and cream when you are re-heating the dish. If frozen dishes do curdle on being thawed this can usually be adjusted by whisking the ingredients in a bowl over a saucepan of hot water but it is preferable to avoid the risk of curdling in the first place.

Salad ingredients: Most salad ingredients will not freeze well in their raw state as they tend to have a high water content. Do not freeze lettuce, unblanched celery or beetroot, radishes, cucumber or spring onions.

Mayonnaise: Mayonnaise will curdle if frozen on its own but you can combine mayonnaise with some chopped ingredients to use as sandwich fillings or incorporate it into jellied moulds where gelatine and other ingredients will stabilise the mayonnaise and prevent it from separating.

6

Storage Times

The length of time you can store food in your freezer
There is still a considerable difference of opinion amongst the experts as to how long you can store and keep food of all kinds in your deep freezer. From the point of view of safety only I think it is probably true to say that almost any foods have an unlimited freezer life providing the freezer maintains a suitably low temperature at all times. From the point of view of colour, flavour and texture, however, there are certain guidelines well worth sticking to in order to avoid disappointments when you finally thaw your raw, blanched or pre-cooked ingredients. Long storage at normal household deep freeze temperatures does not make food dangerous to eat but it can make it tasteless and unappetising and I see no reason why in home deep freezing just as much as in every day home cooking, quality should not be one of the most important criteria.

Unless you are loading your freezer for a long term price war or an emergency of even worse dimensions, there should not, in any case, ever be an argument for the really long term freezing of food in the home deep freeze since the very essence of successful and economical home freezing is based on the quick turnover of frozen food. Ideally your freezer should be used to store food for the most economical length of time with frozen produce being used at the height of its prime with its flavour and texture completely unimpaired.

General guide to the storage of produce in a home freezer

Meat: Meat, even if you buy it in bulk, should not be frozen in a home freezer for longer than a year. Do not buy more meat at one time than your family can consume within a period of about six months and try to keep the packages of meat rotating in your freezer so that you bring those joints which have been in the freezer the longest to the forefront of the freezer.

Fish: Fish tends to have a shorter freezer life than meat and the fresher the taste of the sea or river the better the final flavour of the fish dish you are cooking will be. Whenever possible try to freeze only fish that is absolutely fresh and use it within three months of freezing.

Shellfish: Shellfish tends to have a shorter freezer life than other fish and, whenever possible, I would recommend trying to use your frozen shellfish within a month in order to get the best results.

Poultry and game: Poultry and game freeze well but again it is best to get a reasonably quick turnover in your freezer for the best results. Frozen chickens and other lean poultry can be frozen for at least six months without deterioration but fat birds such as ducks and high smelling game should, ideally, not be kept frozen for much longer than three months.

Vegetables: There is no point in keeping vegetables from one season frozen until the next so that you are, for instance, still eating up last year's frozen runner beans when this year's crop is already maturing in the garden. Once again the guideline with vegetables is to keep the turnover as rapid as possible and to remember that the sooner you eat them the better the quality of the vegetables will be.

Fruit: Many varieties of fruit freeze astonishingly well and fruit such as currants and raspberries will emerge from your freezer after some considerable time tasting almost exactly the same as it did the day it was frozen but again the same maxim applies – there is no point in eating last year's fruit when this year's crop is fresh for the picking in the garden, so keep an eye on the contents of your freezer and avoid hoarding items for so long that you no longer need them.

The following table gives suggested storage times for individual items.

Suggested storage times

1. Baked products

Bread white and brown	Commercially baked	6 months
	Home baked	3 months
	Unbaked	1 month
Cakes	Un-iced and not filled	6 months
	Iced and filled	2 months
	Yeast cakes (i.e. doughnuts etc.)	2–3 months
	Meringues	8 months
	Fruit cakes rich	6 months
	Fruit cakes normal	3 months
Scones and buns	Baked	2 months
	Unbaked	1 month
Biscuits	Uncooked	4 months
Pastry	Unbaked	8 months
	Cooked	3 months
Choux pastry	Unbaked	3 months
	Baked and filled	2 months
	Baked and unfilled	4 months
Sandwiches		3 months
Batters and pancakes	Uncooked	3 months
	Cooked	6 months

2. Dairy foods

Eggs	Lightly beaten whole eggs or separated and lightly beaten eggs with salt or sugar added	8 months
Do not freeze eggs in their shells or hard boiled eggs		
Butter	Salted	3 months
	Unsalted	6 months
Margarine		4 months

Lard		6 months
Cream cheese		3 months
Soft cheese		3 months
Hard cheese	Whole	6 months
	Grated	4 months
Cream, whipped, double or clotted		2 months
Ice cream	Commercial	3 months
	Home made	2 months

3. Fresh meat and meat products

Beef	Large joints on the bone	12 months
	Large joints off the bone	9 months
	Small cuts	8 months
	Steaks	6 months
	Mince, hamburgers etc.	3 months
	Cubed for stews	4 months
	Salt or pickled beef	6 months
Lamb	Large joints	8 months
	Stuffed joints	2–3 months
	Chops, noisettes, cutlets etc.	6 months
	Cubed for stews	4 months
Pork	Large joints	6 months
	Stuffed joints	2 months
	Chops	4 months
	Fat pork, e.g. belly	3 months
	Salt pork	2 months
Veal	Large joints	9 months
	Stuffed joints	2 months
	Escallops and chops	3 months
Smoked meats	Vacuum packed	3 months
	Home packed	2 months
Dripping		2 months
Fat	For rendering down, suet etc.	2 months
Offal	All offal has a shorter freezer life than joints etc.	2 months

A Side of Beef

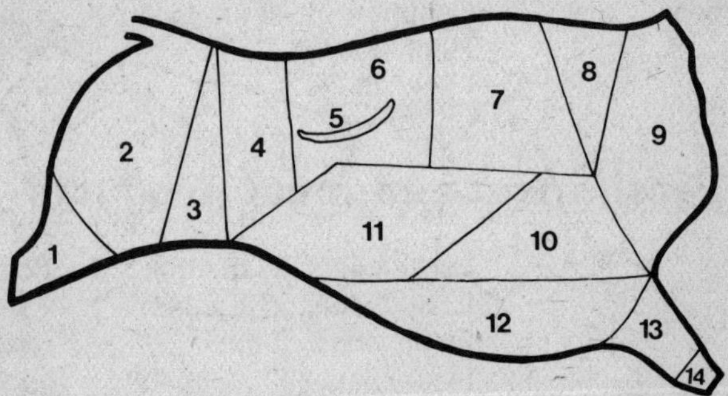

1. Shin
2. Round: round steak and topside; silverside and tail end of rump
3. Aitchbone
4. Tail end of rump; head of sirloin steak
5. Fillet steak
6. Baron including sirloin steak (head end)
7. Rib roast
8. Chuck steak
9. Neck
10. Shoulder
11. Short and rolled rib, flank and skirt
12. Brisket
13. Shin
14. Knuckle

A Side of Lamb

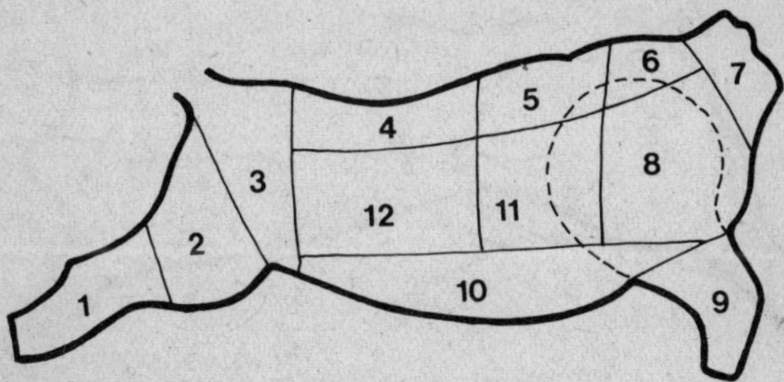

1. Shank
1. and 2. Leg
3. Fillet
4. Loin chops or
4. and 12. Saddle (taken from a whole lamb)
5. Best end (used for Crown Roast and Guards of Honour)
6. Middle neck chops and cutlets
7. Scrag end of neck
6. and 8. Shoulder (off the bone for stewing)
9. Shank
10. Breast
11. Cutlets
12. and 4. Loin chops or saddle (taken from a whole lamb)

Dotted line – whole shoulder of lamb

A Side of Pork

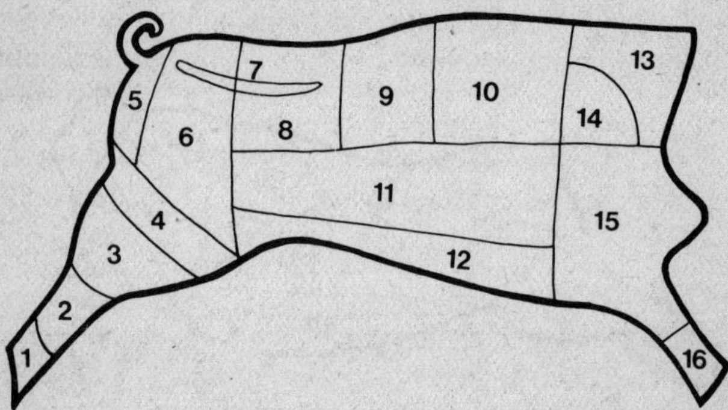

1. Trotter
2. Trotter
3. Knuckle or hock
4. Leg (gammon slices)
5. Slipper

5. and 6. Gammon

7. Fillet
8. Loin (hind), chump chops and loin chops
9. Fore loin and loin chops
10. Chine or shoulder cutlets
11. Streaky bacon
12. Belly
13. Spare rib
14. Blade bone
15. Hand
16. Pig's Trotter

Sausages	Fresh	2 months
	Smoked	3 months
Bacon	Joints, rashers, vacuum packed for freezing	5 months
	Vacuum packed for home use	2–3 months
	Home packed joints	
	smoked	2 months
	unsmoked	6 weeks
	Home packed rashers	
	smoked	6 weeks
	unsmoked	4 weeks

4. Poultry and game

Poultry Pack giblets separately unless buying commercially frozen birds

Chicken	Commercially frozen birds with giblets	9 months
	Home frozen for roasting	9 months
	Joints	6 months
	Off the bone	4 months
	Giblets	2 months
	Whole stuffed	2 months
	(Stuffing frozen separately)	2 months
	In sauce or made-up dishes	2 months
Duck	Whole for roasting	8 months
	Commercially frozen	9 months
	Joints	6 months
	Giblets	2 months
	Whole stuffed	2 months
	In sauce or made-up dishes	2 months
Turkey	Commercially frozen	9 months
	Joints	6 months
	Off the bone	4 months
	Giblets	4 months
	Whole stuffed	2 months
	Stuffing (frozen separately)	2 months
	In sauce or made-up dishes	2 months

Game Plucked and drawn		
Grouse	Young and tender for roasting (protect bones for packing)	6 months
	Old, jointed for casseroling (protect bones for packing)	6 months
Partridge and Pheasant	Young and tender	8 months
	Old, whole or jointed for casseroling (protect bones for packing)	6 months
Pigeon	Whole	8 months
Snipe	Whole (wrap individually in clingfilm and over-wrap)	6 months
Woodcock	Undrawn	4 months
	Drawn	6 months
Hare	Large joints	8 months
	Small joints	6 months
Rabbit	Large joints	8 months
	Small joints	6 months

5. Fish

Much of the success of freezing fresh fish depends on good packaging. Ensure that fish is *really* fresh and in prime condition to avoid disappointments. Freeze fish trimmings for stock making if fish are filletted before freezing.

Whole Fish	Large	6 months
	Small	4 months
Steak or fillets of fish		3 months
Herring and Mackerel whole or filleted		4 months
Salmon and Trout	Whole	4 months
	Steaks and cutlets	4 months
Smoked fish	Whole	6 months
	Sliced	3 months
	Processed (i.e. patés)	2 months

6. Fruit

Apples	Blanched slices	8 months
	Puréed	12 months
Berries and Currants		12 months
Plums, Greengages, Damsons etc.		8 months
Citrus Fruit	Juice and grated rinds	12 months
	Whole	6 months
	Segments	6 months
Grapes		6 months
Melons	Raw or in syrup	8 months
Peaches and Nectarines	In syrup	6 months
Pears	In syrup	6 months
Pineapples	Raw or packed in syrup	6 months
Rhubarb		12 months
Fruit purée	Raw	12 months
	Cooked	12 months
Fruit Salad	(do not add bananas, apples or pears)	6 months
Glacé or crystallised fruit		12 months
Candied Peel		12 months
Leftover tinned fruit	Pack in rigid containers	6 months

7. Pre-cooked foods

Note: These are guidelines only. Most of these foods have longer potential storage times than those given but I do feel very strongly that pre-cooked foods benefit greatly from not being kept for too long in the freezer; the quicker your turnover the better the final results will be.

Label food carefully with freezing and length of storage times and rotate packages so that pre-cooked food is not forgotten in the back of the freezer.

Casseroles: stews, ragouts etc., in brown sauces	4 months
fricassees etc., in white sauces	3 months

Croquettes and rissoles		3 months
Curries		3 months
Deep fried food in batter		2 months
Deep fried food in breadcrumbs		2 months
Dips		1 month
Fish dishes	Shellfish	1 month
	Other fish	2 months
Flan cases	Unbaked	4 months
	Baked blind	6 months
	Filled	2 months
Gelatines, Brawns and other jellied moulds etc.		3 months
Ice Cream	Commercially made	3 months
	Home made	2 months
Meringue puddings i.e. apple meringue, Pavlova etc.		4 months
Minced meat dishes i.e. Moussaka, Cottage and Shepherd's pies etc.		3 months
Mincemeat		4 months
Mince pies		4 months
Mousses	Sweet and Savoury	2 months
Pancakes	Filled	2 months
Pâtés and Terrines	Highly spiced and flavoured	1 month
	Not too highly spiced	2 months
	Uncooked without spices	3 months
Pastry	Cooked	2–4 months
	Uncooked	4 months
Pies	Game or raised	2 months
	Meat	4 months
	Fish	2 months
	Fruit	4 months
	Sweet, general and cheesecake	3 months
Pizzas	Cooked	3 months
	Uncooked	4 months
Potted meats		6 months
Potted fish		3 months
Puddings, general, boiled etc.		3 months
Sauces	White	3 months
	Brown	4 months
	Meat	6 months

Sauces – *cont.*		
	Tomato	6 months
	With eggs and/or cream	1 month
Soups	Vegetable	8 months
	Meat	6 months
	Cream	1 month
	Fish	1 month
Stocks	Vegetable	3 months
	Meat and poultry	6 months
	Cooked carcase	3 months
	Fish	3 months
Stuffings and forcemeats		2 months
Vegetable dishes	Stews and casseroles	6 months
	In sauces	3 months
	Stuffed	2 months
Water ices and sorbets		3 months

7
Thawing Guide

Food that is to be served cold without re-cooking

The more slowly thawing is done the better the results will be. Thaw food that is to be served cold (i.e. savoury mousses, pâtés etc.) in their wrappings in the lowest part of your refrigerator. The time taken will depend on the thickness and consistency of the dish to be thawed but it is usually safer to take the dish out of your freezer the night before you want to use it. Unwrap pastry dishes that are to be served cold before thawing.

Re-heating pre-cooked dishes to serve hot

This is always a tricky process from the timing point of view. One safe way to judge the timing is to remove the dish from the freezer the night before it is required, let it thaw out in a refrigerator overnight or at room temperature for 3–4 hours, and then re-heat it in the normal time it takes to re-heat a made up dish.

Dishes packed in foil containers will re-heat in the oven more quickly than in a fireproof container.

Do allow plenty of time for re-heating frozen dishes. Quite apart from the fact that nothing is more unattractive than a dish which is hot on the outside and still frozen solid in the centre, remember that food that has been frozen is vulnerable to salmonella bacteria. So do ensure that your pre-cooked dishes are heated right through and that the centre as well as the outside reaches boiling point. Test the food by plunging a skewer through to the centre.

Watch pies and puddings with pastry tops to ensure they do not burn while re-heating (often the inside of the dish will need far longer than the pastry crust). Protect the tops of pies with foil or dampened greaseproof paper. It is really better to freeze covered pies with the pastry uncooked.

Soups, stews and casseroles can be heated on top of the stove in a *heavy bottomed* pan over a low heat. If you wish to heat food straight from the freezer, break up the frozen ingredients with a fork as they thaw to spead the heat. As a rough guide most pre-cooked dishes for six servings will need about 1 hour thawing time in a moderately hot oven (375°F, 190°C, Reg. 5). Reduce heat or cover the top of the dish if it begins to brown. Raise the heat for the last 15 minutes if the top of the dish is to be browned.

Using a microwave oven

For thawing or re-heating guides for frozen food in a microwave oven please follow the maker's instructions.

General thawing guide

This is only a general guide as times will vary depending on the temperature of your freezer, your refrigerator and your kitchen. Thawing times will also vary depending on the thickness as well as the weight of food.

Instructions		**Thawing at room temp.**	**Thawing at refrigerator temp.**
Meat			
Large joints	Leave in wrapping	2 hours per lb	5 hours per lb
Small cuts	Leave in wrapping	2 hours per lb	4 hours per lb
Steaks etc. Free frozen	Leave in wrapping	1 hour per lb	3 hours per lb
Poultry	All poultry and game must be completely thawed before cooking		
Chicken	Leave in wrapping	8 hours	Overnight
Chicken pieces	Leave in wrapping	3 hours	6 hours
Duck	Leave in wrapping	8 hours	Overnight

		Thawing at room temp.	**Thawing at refrigerator temp.**
Goose	Leave in wrapping	10 hours	Overnight
Guinea fowl	Leave in wrapping	3 hours	8 hours
Turkey	Leave in wrapping		
	5–6 lb	24 hours	50 hours
	6–8 lb	30 hours	60 hours
	8–10 lb	36 hours	66 hours
	10–12lb	42 hours	70 hours
	12–15 lb	48 hours	74 hours
	15–17 lb	54 hours	78 hours
	17–20 lb	60 hours	82 hours
	20–25 lb	66 hours	86 hours
Game	Leave in wrappings		
	Large birds	4 hours	Overnight
	Small birds	2 hours	8 hours
Venison	Leave in wrappings		
	Large joints	2 hours per lb	5 hours per lb
	Small cuts	1 hour per lb	2 hours per lb
Fish	Leave in wrappings thaw only in refrigerator		
	Large fish		24 hours
	Flat fish		6–8 hours
	Fish fillets	Cook frozen or thaw in refrigerator	4 hours
	Breadcrumbed or battered fillets. Cook frozen. Thaw only in the refrigerator.		8 hours
Vegetables	Most vegetables should be cooked straight from the freezer in very fast boiling water.		
Fruit			
To eat as fresh:	Thaw only in the refrigerator serve while still slightly frozen		6–8 hours
To cook	No need to thaw		
Purées, raw		2–3 hours	6–8 hours

		Thawing at room temp.	**Thawing at refrigerator temp.**
Dairy Foods			
Eggs		2 hours	4 hours
Butter		3 hours	6 hours
Margarine and lard		2 hours	6 hours
Cheese	Hard	2 hours	6 hours
	Soft	1 hour	5 hours
Cream double or clotted		2 hours	6 hours
Bread and cakes			
Unbaked cakes	Thaw in wrapping	2 hours	5 hours
Baked	Thaw in wrapping	2–4 hours	4–8 hours
Unbaked bread dough	Thaw in wrapping	2–4 hours	4–8 hours
Baked bread (sliced bread can be toasted from the freezer)	Thaw in wrapping	2 hours	4 hours
Unbaked pastry	Thaw in wrapping	1 hour	4 hours
Baked pastry	Unwrap	2 hours	5 hours
Sandwiches	Thaw in wrapping	4 hours	8 hours
Batter		2 hours	4 hours
Pre-cooked foods			
Casseroles, soups and stews		4 hours	8 hours or overnight
It is not always necessary to thaw before re-heating			
Cooked sliced meats		4 hours	8 hours
Sauces	Hot	Not usually necessary to thaw before using	
	Cold	4 hours	8 hours
Cold sweet and savoury mousses, soufflés and other jellied foods		Thaw in the refrigerator	8 hours or overnight

8
Packaging for the Freezer

Packaging for the freezer

Good packaging is one of the most important elements of successful freezing. Inadequately packaged produce deteriorates quickly and suffers loss of flavour and texture in a very short time. Badly packaged items quickly develop 'freezer burn' resulting in discolouration and the produce becoming dry and tough with a high loss of fluid when the food is thawed.

Deep freeze packages must be strong enough to stand the rigours of being bumped and moved around in the freezer and must be completely airtight so that ice crystals will not form inside the package once it has been frozen.

Packaging materials must be moisture and vapour proof and there are now a wide variety of packaging materials from which to choose. These materials are not cheap and can add greatly to the expense of your freezer budget so, rather than buying them in small quantities from a stationers, buy in bulk from a manufacturer specialising in equipment for the freezer. The initial outlay may be quite expensive but you will save in the long run and the packaging materials themselves have a long life if they are stored in a cool, dark place. Like everything else packaging materials are subject to inflation so the more you buy now the more you will save in the long run.

Rigid containers have an advantage over other packaging materials as they are easy to label, either with stick-on labels or with a Chinagraph or waterproof felt pen straight onto the box – stick on labels being preferable as waterproof markings are almost impossible to remove when the box is to be re-used.

Some rigid polythene or plastic containers have coloured lids for colour coding.

Note: Empty rigid containers that are to be stored, packed one inside the other can sometimes stick together. To separate, dip the containers into hot, soapy water, rinse and dry well.

1. Packaging materials and equipment and how to look after them

Polythene bags: Probably the most popular of all freezing materials for the home freezer and a field in which there is such a wide and ever increasing variety of strengths, shapes, sizes and colours to pick from that the choice can be a confusing one. Do not attempt to economise by buying inferior or inexpensive bags; it is not worth while in the long run. Buy bags that are made especially for freezing and are strong enough to last a long time and can be used again.

Choose the right bag for the product you are freezing. The neater the fit the easier it will be to press out all excess air. It is a waste of money to pack small items in overlarge bags and bad freezing to try and force produce into bags that are too small; since bags come in all shapes and sizes it should always be possible to find the right bag for the job whether you require a long thin one for a salmon, a large gusseted bag for a turkey or a mini bag for herbs. Always keep a large supply of assorted polythene bags at hand.

Plain polythene bags: These come in every shape and size. Make sure those you buy are made of heavy-duty polythene, suitable for freezing, if you plan to freeze produce packed in them for any length of time.

Plain polythene bags with coding colours: Many of the freezing manufacturers produce bags that have stripes or bars of colour on them for easy identification and these can be especially useful if you plan to freeze fruit and vegetables on a large scale. Some polythene bags have a white area on them for labelling which I find preferable to tie-on or stick-on labels which can get lost.

Extra heavy-duty bags for special purposes: The stronger the polythene the longer the protection it will afford for the produce inside the bag. Heavy-duty polythene bags should be used for any produce liable to pierce a less strong bag, i.e. fish with spines, poultry, game or meat with exposed bones. (*Note*: exposed bones or other potentially sharp objects should be given extra protection by being wrapped before they are sealed in a polythene bag.)

Gusseted bags: These are bags with a built-in pleat which makes it easier to get a neatly shaped package when packing fish, meat and poultry etc.

Self sealing bags: Bags with self sealing grips rather like a smooth zip that you press together. Small self sealing bags are useful for items such as herbs, lemon peel, grated cheese etc. but the seals are not really reliable enough for long term freezing.

Plastic bagging by the yard: Continuous bagging (sealed on both sides) which can be cut to the size you require. The open ends of the bags should be sealed with a special sealing iron (*see* below).

Boiling bags: Heat treated bags which can be taken straight from the freezer and dropped into boiling water without disintegrating. The bags can be used for small amounts of vegetables, individual portions of stew, curry etc. but the quantities should be small or the ingredients on the outside of the package will be cooked while the inside remains frozen.

Re-cycling and care of polythene bags: Do not throw bags away lightly. In a great many cases polythene bags can be used again and again (protect bags from the odours of fish or other strongly smelling produce by wrapping the produce in transparent wrapping film before bagging).

Wash bags by turning them inside out, rinsing them in water to which a little detergent has been added, rinsing them well in clean water, drying them inside and out with a tea towel and then pegging them on a washing line or draping them over a bottle to allow them to dry out completely. Store the washed bags in a cool, dry, dark place.

Transparent polythene wrapping film: Thin polythene sheeting which seals itself when the edges are pressed together. Like foil the film comes in rolls and the larger the roll the cheaper will be the film. This is a very useful product for wrapping ingredients to give a really airtight package and especially for wrapping items like sandwiches and cakes. The film is not strong enough for actual freezing so ingredients wrapped in it should always be overpacked in polythene bags. The film can also be used to separate items that are to be packaged together thereby preventing them from sticking, i.e. fillets of fish, chops, steaks etc. Wrapping whole fish in the film helps to keep it moist and in good condition during its freezer life.

Rigid polythene containers: Excluding air is the great problem when freezing at home and although polythene containers probably provide one of the most convenient and easily stored methods of packaging for the freezer they can only really be used for ingredients that are, to a certain degree, liquid and will freeze in a solid mass. Other ingredients need to have air spaces packed with crumpled kitchen paper in order to cut down on as many potential pockets of air surrounding the food as possible.

If the ingredients are liquid, room for expansion is necessary or the lid will be forced off as the liquid expands on freezing (this applies especially to soups, sauces, stews, casseroles, purées etc.).

Ingredients must be cooled down before they are packed and sealed in polythene containers.

Rigid polythene containers come in all sizes and shapes and also in a considerable variety of strengths and again it is worth buying the best. Good quality containers can be used over and over again. Wash them in detergent, rinse well and wipe completely dry before re-using.

Once the ingredients to be sealed have been packed in the container the lid should be snapped on firmly. Lift one corner slightly, press the centre of the box to expel any air and then snap the lid on firmly again.

Sealed containers are suitable for short term freezing as they are, but if you intend keeping the ingredients in the freezer for some time it is well worth over-sealing with freezer tape or packing the container in a sealed polythene bag just in case the box opens.

Foil and foil containers

Kitchen foil: Bought by the roll and the longer the roll the better the value. Special heavy-duty foil is recommended for wrapping items to be frozen as it provides added protection and prevents air damaging the texture of the food. But, as foil cannot be sealed with 100% reliability, foil wrapped packages should be over-wrapped in polythene bags. Items to be frozen for only a short time can be double wrapped in foil and sealed with freezer tape.

Foil bags: Bags of strengthened foil which can be used for both freezing and cooking from the freezer. Packages must be carefully sealed and cannot usually be used again so this is a rather expensive form of packaging material.

Foil containers: Very convenient for storing stews, casseroles and pies etc. since the ingredients can be cooked, frozen and re-heated in the containers. Foil containers are often sold with lids but since these do not provide an adequate seal to protect the ingredients for anything but a short time the containers should usually be overwrapped in a polythene bag before being frozen. The containers come in a large variety of shapes and sizes and there are some quite attractively patterned circular foil dishes which can be used for freezing and for serving food at the table.

Waxed cardboard cartons and containers: I would not really recommend these for very long term freezer use and although they are less expensive than rigid polythene containers, they cannot be washed and re-used, and so may be less economical.

Do not pour hot liquid or hot ingredients into waxed containers for obvious reasons.

Strengthened waxed paper bags with polythene linings: The bags are provided with a built-in wire closure and are sealed by folding over the top of the bag and then securing the fold by turning in the ends of the wire. Do not fill the bags with hot ingredients.

Strengthened glass dishes: Some glass dishes are made in specially strengthed glass which will stand both freezing and re-heating. These are useful for the short term freezing of stews and casseroles etc. but since they are expensive you will obviously not want to have them out of action in the freezer for too long. Allow a good half inch headroom for expansion and overwrap in a polythene bag.

Glass bottles or jars that are not strengthened are not recommended for freezing.

Waxed cardboard: Fragile items like pies, flans, tarts and cakes need to be protected so they will not get damaged. It is best to free freeze these items first, without packing and then to pack them when they are frozen solid but they also benefit from the protection of a casing of waxed cardboard before being packaged. These fragile dishes can also be packed in rigid containers provided all the air spaces are filled with crumpled kitchen paper to exclude as much air as possible.

Waxed freezer paper: Used for separating items in a package i.e. buns or biscuits that are packed in a rigid container or plastic bags – separating the items makes thawing much more easy.

Free packaging materials: Do not throw away undamaged yoghurt, cream and ice cream containers etc. Plastic containers can be re-used provided they are carefully washed out and dried before being used. The lids will usually have been damaged so cover the top of the filled container with transparent wrapping film and then overwrap in a polythene bag. Small cartons are very useful for storing small quantities of leftover gravy, sauces, concentrated stocks and egg yolks or whites.

2. Sealing materials

For successful freezing the packaging in which produce is to be stored must be carefully sealed to make sure no air is allowed to enter the package during its freezer life.

Freezer tape: Use to seal packages for short term freezing that are wrapped in foil, waxed paper or transparent wrapping film and for sealing waxed, polythene, or plastic containers (many of these are 'self sealing' but it is worth oversealing with freezer tape as the lids *can* open in the freezer).

Heat sealing machines or irons for sealing polythene bags: There are now quite a range of these gadgets available and many of them are sold complete with special lengths of polythene bagging which you seal off in the sizes you require. The sealing has to be done very carefully so that no air can be allowed to enter the bag and it is important to expel as much air as is possible before the bags are sealed. Most machines have to be kept very clean and free of dust or dirt as they may not function properly.

Wire ties: Paper and plastic covered lengths of wire can be bought in bulk (the plastic covered variety also come in different colours for colour coding). Plastic covered ties can be re-used.

Paper tags can also be used for sealing foil and polythene lined paper bags. Place the wire along the top of the bag about quarter inch down from the top, fold over the top of the bag a couple of times and then fold the ends of the paper tag firmly inwards to ensure a tight seal.

Rubber bands: Strong rubber bands can be used in the place of wire tags and can be re-used. Twist the band two or three times around the top of the bag as you expel the air from the package.

Coloured plastic tags with notches: Used to seal plastic bags but they do not give a really 100% seal.

3. Labelling materials

There is no point in taking the trouble to package carefully if you then do not label the produce correctly. If you are freezing a large amount of produce, packing it in coloured polythene bags or in rigid containers with coloured tops helps you to see at a glance what category the contents fall into but all packets still need to be clearly labelled with the contents, the date when they were frozen, the date by which they should be used and the quantity or number of servings.

Labels

Paper Labels attached to wire ties: Should not be used for long term freezing as the labels tend to disintegrate after a time.

Self sticking labels: These come in different colours for coding but although they are efficient they still sometimes come off a package if it has been left for some considerable time in the freezer. If you plan to freeze for more than a few months the labels should be placed inside the polythene bags. Do not try to write on labels that are too small; the labels must be large enough to contain all the information that is required.

Chinagraph pencils: These come in different colours for coding and can be used to write on almost any packaging material. The Chinagraph pencils are throughly waterproof and it is almost impossible to remove the writing once the package has been used; if the pencils are to be used for re-usable rigid containers it is worth sticking some tape over them and labelling that rather than the container itself.

Waterproof felt pens: Used in the same way as the Chinagraph pencils (see above).

Correct labelling

Contents	Quantity	Freezing date	Use By
Beef and onion stew	Serves 4	7.9.77.	7.12.77

Keeping a record

When you put produce into the freezer there seems to be no possibility of forgetting what you have stored there but within a week you will have forgotten what the freezer contains and the packages will have become unrecognisable soon after the produce is frozen solid and a certain amount of icing has inevitably occurred inside the package. Keep a record book (freezer packaging material firms produce special books for this purpose) and record all produce as it goes in, and as it comes out of the freezer.

4. Other useful equipment for freezing

All the following items can be bought from a good deep freezing equipment firm.

A wire basket for blanching: If you are planning to freeze a lot of vegetables a large wire basket (you can use a deep-frying basket) which will fit into a large saucepan will be a great help for blanching. The prepared produce is put into the basket and the basket then lowered into fast boiling water. The water should return to the boil and the blanching time is estimated from the moment the water is bubbling. The ingredients are easily lifted from the saucepan in the basket and the same water can be re-used for the next lot of blanching.

Plastic trays for free freezing: Many items benefit from being free frozen without packaging until solid and then being packed and sealed. Lightweight plastic trays are used to spread the ingredients on and will help to speed up the freezing process.

Plastic carrier bags: If you have a chest type freezer the plastic bags you get from your grocer are very useful for packing together individually frozen items, properly packaged, to make it easier to find them, i.e. pack home frozen blackcurrants in their bags, inside a carrier bag and mark the bag with a Chinagraph pencil or waterproof felt pen.

Extra wire trays or partitions: Freezers are supplied with trays or wire separators but it is possible to buy more should you need them.

5. Hints on packaging

Filling polythene bags to form square or oblong shapes: If you are packaging fairly liquid ingredients in polythene bags it is useful to form as regular a shape as possible. The bag can be put into an empty carton (I use something like an empty sugar packet), frozen and then, when the ingredients are solid, the box is removed

Alternatively the ingredients can be frozen in a square or oblong container until solid, dipped in hot water, unmoulded and then packed in a polythene bag.

Filling plastic bags with produce: This can be a fiddly process and is made easier if the bag is placed inside a wide topped jar or empty carton. Fold the top of the bag over the top of the container, fill the bag with produce to the top of the container, gather up the edges, seal and remove the filled bag from the container. The filled bag can be frozen in a rectangular carton to make for easier storage.

Prawns, scampi etc.: Free freeze by spreading on trays and freezing until solid before wrapping. Package in a suitable container as soon as the fish is frozen.

Berries: Free freeze without packaging, spreading the berries out on a tray and freezing until they are solid before packaging in a suitable container.

Beans and peas: Free freeze without packaging until the beans are frozen through and then package in a suitable container.

Liquid ingredients: Always leave about half inch headroom in all packages containing liquid ingredients as these will tend to rise slightly when frozen and can force the lids off rigid containers. Solid ingredients, such as meat, poultry and fish, should be packaged as tightly as possible.

9

Preparing and Freezing Meat, Poultry, Fish and Game

1. Preparing and freezing meat at home

Do not try to freeze more than one-tenth of the total capacity of your freezer at one time or the freezing process will be slowed up and the results not as good as they should be.

Always prepare meat for the freezer in scrupulously clean surroundings. Divide meat into joints and trim for cooking before freezing. Cut off excess fat before freezing to prevent rancidity.

Bones take up an unnecessary amount of room in the freezer so cut them out whenever possible and use them to make stock.

Meat to be stored for any length of time should be very carefully packed indeed in heavy duty packaging material. Meat is highly susceptible to freezer burn so it needs good protection and airtight packing with as much air as is humanly possible squeezed out of the packages.

Large joints of meat: Remove bones when possible. Wipe meat with a clean, damp cloth and prepare ready for the oven (i.e. roll and tie etc.). Leave on enough fat to keep the joint moist during cooking but cut and trim off all excess fat.

Wrap the ends of awkward or sharp bones with foil or wrap the whole joint in cotton stockinette to protect the outer packaging.

Pack joints in heavy-duty polythene bags that are the right size for the job. Press out all excess air, seal, label and freeze with the fast freeze switch turned on.

Small cuts: Wipe meat with a clean, damp cloth and prepare for cooking. Pack in small quantities in heavy-duty polythene bags that are the right size for the job. Press out all excess air, seal, label and freeze.

Stewing meat: Cut off all excess fat (leaving on enough to keep the meat moist while cooking) and cut into neat cubes. To save room in the freezer the stewing meat can be packed tightly into rigid containers, frozen until solid and then repacked as soon as it is frozen into heavy-duty polythene bags.

Do not freeze the meat in too large quantities, it is easier to use two bags than it is to try and saw off half a quantity of frozen meat.

Marinate the meat in oil and lemon juice if you think it will be tough.

Minced meat: Mince meat before freezing rather than mincing meat that has been frozen and thawed – in this way you will loose as little juice as possible. Pack minced meat in small quantities in heavy-duty polythene bags or block freeze in small amounts before repacking.

Steaks, chops and other cuts for frying: Individually underwrap all small cuts of meat in transparent wrapping film before packing in heavy-duty polythene bags. Do not pack in too large quantities at a time and express as much air as possible from the bags. Seal, label and freeze.

Offal: Offal must be 100% fresh for freezing. Prepare the offal for cooking and pack in small quantities in heavy-duty polythene bags. Slices of liver can be inner wrapped individually with transparent wrapping film to make handling easier.

Hamburgers, meat balls and other meat products: Prepare for cooking and open freeze before wrapping individually and over-packing in heavy-duty polythene bags. The products can be packed in a rigid container but all air spaces should be filled with crumpled kitchen paper. Square items are easier to package.

Sausages: Free freeze for convenience before packing in small quantities in heavy-duty polythene bags. Check storage times.

Sausage meat: Pack in small quantities and put in heavy-duty polythene bags. Press out as much air as possible. Seal, label and freeze.

Salted or cured meat: Do not soak before freezing. Wipe over with a dry cloth removing all excess moisture. Pack in heavy-duty polythene bags. Press out as much air as possible. Seal, label and freeze.

Bacon and ham products: Bacon and ham must be absolutely fresh and in top condition if you want to freeze it. Avoid produce that is slimey or has been sitting in the sun. Smoked bacon has a longer storage life than green bacon but large joints cannot be frozen quickly enough in a home freezer to keep within the safety level so only freeze joints that are under 5 lb in weight.

All ham and bacon joints and rashers should be cooked as soon as possible after thawing.

Bacon joints: Wipe with a clean, damp cloth, prepare for cooking and underwrap with cotton stockinette or foil for extra protection. Overwrap in heavy-duty polythene bags, press out all excess air. Seal, label and freeze.

Bacon rashers: Pack in heavy-duty polythene bags, press out all excess air. Seal, label and freeze.

Commercial packs: Make sure packets have not been damaged or pierced in any way. Label and freeze.

Gammon rashers or steaks: Divide steaks or rashers with freezer paper or transparent wrapping film, overwrap in heavy-duty polythene bags, press out all excess air. Seal, label and freeze.

Ham: *In one piece:* Wrap in transparent wrapping film to protect surfaces from freezer burn. Overwrap in heavy-duty polythene bags. Press out all excess air. Seal, label and freeze.

In slices: Interleave slices with transparent wrapping film for convenience after thawing, overwrap in transparent wrapping film and pack in heavy-duty polythene bags. Press out all excess air. Seal, label and freeze.
Note: Pack in small quantities.

2. Preparing and freezing poultry at home

Remove the giblets and freeze separately. (You may, if you are freezing a number of birds at one time, like to divide the giblets into livers, hearts, gizzards etc. to use for specific purposes.)

Chicken or other poultry portions: Free freeze or individually wrap portions in transparent wrapping film before overwrapping in heavy-duty polythene bags. (Giblets can be packed in smaller bags and frozen with the bird but remember they have a shorter freezing life.) Press out all excess air, seal, label and freeze.

Whole birds: Prepare for cooking. Wipe all over the outside and inside with a clean, damp cloth. Protect any sharp bones with foil. Pack in gusseted, heavy-duty polythene bags. Press out all excess air, seal, label and freeze.

3. Preparing and freezing game at home

Hang all game before freezing.

Pheasant, partridge, pigeon, grouse: Hang birds head down in a cool, dry place, from 4 to 8 days depending on how high you like your game and on weather conditions prevailing during the hanging time. Pluck the hung birds; remove the heads but leave on the feet; pick out any shot if possible, and clean out the insides of the birds. Wrap the feet and wing-tips of the birds in foil to prevent any perforation of the packaging material during freezing; pack in heavy-duty polythene bags; press out all the air; seal, label and freeze. The long tail-feathers of pheasants can be frozen with the bird for decoration, but they should be carefully wrapped in foil to prevent their being damaged. Tough old birds can be jointed before freezing to use for casseroles and made-up dishes – pack as for chicken portions.

For pâtés and terrines, strip the flesh off the birds (use the bones to make a rich stock). Pack the flesh in heavy-duty polythene bags.

Woodcock and snipe: Hang the birds head down for 4 to 8 days in a cool, dry place, depending on how high you like your game and on the weather conditions prevailing. Pluck carefully leaving the head, feet and entrails intact. Twist the head back and use the long beak as a skewer to anchor the legs in a trussed position. Wrap the birds in two layers of foil to prevent perforation of packaging material by the sharp beaks, and pack in heavy-duty polythene bags. Press out all air, seal, label and freeze.

Venison: Venison must be hung before freezing or the meat will be tough and stringy. It should be hung for 5 to 8 days, depending on weather conditions. The meat is usually sold in cuts similar to those of beef and should be packed in the same way. Any tough cuts can be cut up for casseroling and frozen with a marinade of oil and lemon juice to help tenderise the meat.

Hares: Hang hares head down for 5 to 6 days in a cold, dry place (hares should not be cleaned and drawn before being hung). Attach a bowl to the head to catch any blood (this helps to make a rich gravy for jugged hare and other dishes). Skin and clean the hare, reserving the liver. Divide into joints (unless the hare is to be roasted whole) and wipe well with a clean, damp cloth. Wrap each joint in transparent wrapping film, pack with the liver in a heavy-duty polythene bag; press out all air; seal, label and freeze. The blood can be frozen in ice-cube moulds and packed with the joints.

If you have more than one hare the saddles can be packed and frozen separately for roasting, and the remaining joints packed for stewing or jugging. The joints of a tough, old hare can be marinated in oil and lemon juice before freezing to help tenderise the meat.

Rabbits: Rabbits should not be hung. Skin and clean and divide into joints. Wipe the joints with a clean, damp cloth, wrap in transparent wrapping film and pack in a heavy-duty polythene bag. Press out all air; seal, label and freeze.

4. Preparing and freezing fish at home

Because of the strong smell of fish and the sensitive nature of its texture and flavour extra special care should be taken with packing.

Large fish, frozen whole: Prepare for cooking or leave whole. Wipe the fish with a clean, damp cloth, dip fish into cold water and open freeze until a solid skin of ice has formed over the fish (for long term storage the fish can be dipped again once it has frozen to make a stronger protective layer of solid ice). For short term freezing there is no need to cover the fish with the film of ice.

Wrap the fish in a layer of transparent wrapping film and then overwrap in a gusseted, heavy-duty polythene bag. Press out all excess air, seal, label and freeze making sure your label states whether or not the fish was gutted and cleaned before freezing.

Small fish, frozen whole: Prepare for cooking, free freeze and pack in heavy-duty polythene bags. Seal, label and freeze.

Fillets or steaks: Prepare for cooking by skinning and boning etc. Rub oily fish with a little lemon juice and wrap fillets or steaks individually in transparent wrapping film before overwrapping in heavy-duty polythene bags. (Fillets and steaks can be free frozen before packaging.) Press out all excess air; seal, label and freeze.

10

Preparing and Freezing Vegetables and Herbs

Growing your own produce for the freezer

You must surely have friends who spend most of their summer picking, podding, shelling, blanching and freezing the produce from their gardens and never have time to go to the seaside or merely to sit and enjoy all that hard work they have put into their gardens. It is never worthwhile becoming a slave to anything and with careful and sensible planning it should be possible both to benefit, through your freezer, from your garden produce and to have a relaxed summer as well. The well-planned garden should produce vegetables all the year round and if you have a reasonably large kitchen garden or an allotment there is no point in taking up valuable space in your freezer by freezing quantities of vegetables you will never eat because there is a variety of fresh produce available instead. I have actually seen people spending hour upon hour of their time freezing everything they grow, never eating any vegetables straight from the garden when they are at their absolute best but instead eating the beans they froze in the summer during the winter months and the cabbages and Brussels sprouts they froze in the winter during the summer.

Plan your garden for eating fresh and freezing so that you get a good balance of produce. Freeze any gluts that may occur but do not grow vast quantities of peas and runner beans just for freezing – these items are produced on such a large and mechanised scale by freezer companies that, in the long run, you would probably do better to buy them in bulk from a deep freeze supermarket anyway.

Do grow and freeze vegetables your family is particularly fond of and which may be expensive to buy in fresh or commercially frozen forms; grow and freeze tomatoes, aubergines, peppers and sweetcorn, also luxury produce like *mange tout* or sugar peas, french beans to pick when they are tiny and crisp, mini courgettes which make a good vegetable at any time of the year and soft fruit which freezes well and which will keep you in puddings throughout the winter months.

Do not be tempted to cheat on the preparation procedure of fresh vegetables and fruit because you suddenly have a massive crop on your hands – in the long run you will only be wasting time as produce which is not prepared properly and packaged carefully for freezing will only give inferior and disappointing results.

Pick your produce in quantities which you will be able to deal with quickly as the quality begins to deteriorate as soon as the produce is picked. Always pick first thing in the morning if possible, as soon as the dew has dried and before the sun has reached full strength. Resist the temptaion to prepare your fruit for freezing out in the sunshine; it may be far more pleasant to sit in the garden to shell and pod your produce but the heat will suck out a lot of the flavour.

Process and freeze all produce as quickly as possible – the less time it takes you to get your produce from the garden to the freezer the better will be the flavour.

Some garden produce can be frozen without blanching (*see* pages 76–8) but vegetables frozen by this method must be used quickly and are not suitable for long term freezing.

1. Vegetables

As a considerable amount of space in your freezer will inevitably be taken up by frozen vegetables it is essential to ensure you are not risking any loss so prepare the vegetables with care and pack them as professionally as possible for long term storage.

For the successful long term freezing of vegetables blanching them first is essential, to arrest the development of enzymes working in the produce and to guarantee tender results with good colour. Once blanched the vegetables should be cooled as quickly as possible, dried carefully and packed and frozen in as short a time as is possible from the moment they were picked.

Try to keep as wide a variety of vegetables as you can in your freezer, one gets just as bored with the monotony of runner beans with every meal as one would with eating hamburgers every day.

Some of the more tender vegetables respond best of all to being lightly sautéed in butter instead of being blanched before freezing – remember, however, that butter does not have all that long a freezer life so the life of the vegetables is also cut down.

In the following chart the blanching time is for average sized vegetables. If, like me, you are fortunate to be able to grow your own you may like to pick the vegetables when they are far smaller and more tender than those usually sold in the shops. These small vegetables will obviously need a slightly shorter blanching time.

Do not skimp on the preparation of vegetables before freezing – those strings on the sides of beans are not going to get any less tough in the freezer. Try to grade vegetables into equal sizes so that they will blanch and cook evenly. **Freeze only top quality vegetables for the best results.**

Dealing with vegetable gluts

If you grow your own vegetables there are times when, however well you have planned and staggered your planting, everything seems to come at once and, due to weather factors, you find yourself producing far more of one variety than you can really cope with. Letting the vegetables rot in the garden is a nightmare but so is working your fingers to the bone night and day to get the produce frozen, only to find in the long run that much of it had gone over the top and has not really frozen satisfactorily or that at the end of the freezing year, you are left with quantities of frozen vegetables which will have to be sacrificed to make way for the new season's crop. If you do have a glut try to see if a neighbour is prepared to swop some of your excess harvest for some of his of another variety.

Do not attempt to freeze vegetables which have gone over the top as the effort will just not be worthwhile. If you have allowed vegetables to get past their prime and toughen, cook and purée them before freezing or turn them into nourishing and inexpensive vegetable stews or soups to have during the winter. Old, tough vegetables can also be used to make very valuable stocks with the assistance of a few stock cubes.

Blanching and freezing vegetables

Using the correct equipment when you are dealing with large

quantities of vegetables helps enormously to cut down the time you will need.

Use a large heavy bottomed pan that will boil quickly. Invest in a blanching basket (a wire basket with small mesh to contain the prepared vegetables that can be immersed in the water). Have plenty of ice cubes ready to cool the water into which the vegetables are plunged immediately after blanching.

Do not make the mistake of trying to cope with too many vegetables at one time. Picking a little every morning and processing them quickly is far better than picking every few days and having to cope with a mountain of vegetables to prepare, blanch, pack and freeze. Blanch in small quantities so that the boiling water is able to flow freely through the vegetables. Do not add the vegetables to the blanching water until it is really bubbling fast and time the blanching process from the moment the water returns to boiling point.

Vegetables that discolour when cut and trimmed should have lemon juice added to the blanching water.

Unless the vegetables to be processed have an exceptionally strong flavour like cauliflower, the same water can be used time and time again for different varieties of vegetables although some of it will obviously evaporate as it boils and the pan may have to be topped up from time to time.

Spread out kitchen towels over your draining board on which to spread the cool vegetables so that they dry quickly. I also speed up this process by blowing cold air on the vegetables with a hair drier!

Free freeze most vegetables before packing to keep them separate in the packages and therefore enable them to cook more quickly than if they were in a solid mass. Spread out the vegetables in a single layer and free freeze with the fast freeze switch on until the vegetables are solid. Pack them in polythene bags, seal, label and return immediately to the freezer.

Do not ignore 'side' crops if you grow your own vegetables. Beetroot and turnip tops, for instance are delicious in their own right. Brussels sprout tops make a good vegetable and asparagus stems can be boiled to use for an asparagus soup.

Remember that if you grow your own vegetables most roots will store well in sand throughout the winter so take this into account before taking up valuable freezing space. This of course does not apply to the freezing of *young* vegetables.

Blanching chart for vegetables, cooking times and cooking ideas

	Storage time	*Preparation*	*Blanching time*	*Cooking remarks*	*Cooking time*
Artichokes (globe) whole	12 months.	Remove outer leaves. Trim off tops of leaves with scissors and rub cut leaves and stalks with lemon juice. Invert artichokes to drain well after cooking. Pack in individual polythene bags.	6–8 mins.	Cook from frozen, until leaves will pluck easily from the artichokes. Test the tenderness of the 'fond' by plunging in a carving fork, or skewer.	5–10 mins.
Artichokes (globe) hearts	6 months.	Choose small, tender artichokes. Cut off outside leaves leaving only inside yellow/white leaves. Blanch in water to which 1 tabs. lemon juice has been added.	5 mins.	Use as an hors d'oeuvres in a vinaigrette dressing or add to stews or casseroles	5–8 mins.
Asparagus	9 months.	Buy in season. Trim off tough ends of stems. Handle with great care and pack in rigid containers to prevent the tips being damaged.	2–4 mins. depending on size.	Stems can be used for making soup in water in which the asparagus was blanched. Tie blanched asparagus into small bundles after blanching. Cook straight from freezer.	3–5 mins.

Aubergines	9 months.	Well worth growing in a cold frame. Peel and slice or leave unpeeled and slice or dice. Work quickly to prevent the aubergines discolouring. Free pack slices.	3 mins.	Very useful to add to stews and casseroles to stretch meat. Combine with tomatoes, onions, peppers and courgettes to make ratatouille to serve hot or cold. Cook in boiling, salted water or in oil.	8–10 mins.
Beans, Broad	6–12 months.	Large beans: Pod and grade into sizes. Beans under 6 in long: Leave whole if very tender or slice through like runner beans. Blanch the whole beans in water to which some lemon juice has been added. Free pack beans.	2–3 mins.	Cook in very little boiling, salted water. Drain and toss in melted butter. Large beans may have to be peeled after cooking. Cold peeled broad beans make an excellent salad ingredient	5–8 mins.
Beans, French	12 months. 2 months unblanched	Use only young, tender beans. Top and tail and grade if necessary.	2–3 mins.	Cook in very little boiling, salted water or cook in melted butter without water until just tender but still crunchy.	5–7 mins.
Beans, runner	6 months.	Probably the most prolific croppers of the beans grown in this country. Top and tail, remove any strings from the sides and slice diagonally into ½ in thick slices.	2 mins.	A useful standby which can be given a new lift by the recipes on pages 155, 157. Cook in very little fast boiling water until just tender or in melted butter.	5–8 mins.

	Storage time	*Preparation*	*Blanching time*	*Cooking remarks*	*Cooking time*
Beetroot	6–8 months	Cook beetroots before freezing. Cook small beetroots in their skins until tender. Remove the skins, cool and freeze. Large beetroots should be cooked whole, skinned and then sliced before freezing. Free freeze slices.		Thaw frozen beetroot in the refrigerator. Serve in vinegar with salads or in a cheese or sweet and sour sauce as a vegetable.	1–2 hours
Beet Tops	12 months.	A valuable vegetable rather like spinach which is often overlooked. Wash the leaves, remove the tough stalks and use as spinach.	2 mins.	Plunge frozen beet tops into very little salted water and cook over a high heat until just tender. Use cooking liquid for vegetable soups.	5 mins.
Broccoli	12 months.	Remove outer leaves and trim off tough stems. Free freeze and pack in rigid containers or polythene bags.	1–2 mins.	Cook straight from the freezer in a little fast boiling, salted water. Do not overcook, stems should still be crisp. Serve with melted butter or a sauce.	5–7 mins.
Brussels sprouts	12 months.	Trim off tough outer leaves and stalks and grade into equal sizes. It is better not to cut through the bottom of the sprouts. Free freeze and pack in polythene bags.	1½–3 mins.	Cook straight from the freezer in a little fast boiling salted water. Or cook small quantities in melted butter and season with salt and freshly ground black pepper.	5–7 mins.

Calabrese	12 months.	A delicious green form of sprouting broccoli.	1–2 mins.	As broccoli	
Carrots	6 months unblanched. 12 months blanched.	Scrub small carrots leave them whole. Peel and slice large carrots or cut them into dice for stews etc. Free freeze whole or cut carrots before packing in polythene bags.	3–4 mins.	Add to stews or cook in boiling salted water straight from the freezer.	5–8 mins.
Cauliflower	6 months.	Choose only white, unblemished vegetables. Remove stalks and outer leaves and divide into florettes (very small caulis. can be left whole). Free freeze florettes and pack in polythene bags.	3 mins.	Plunge into boiling, salted water straight from the freezer. Do not overcook.	5–7 mins.
Celery	4 months.	Must be blanched and used for cooked dishes only because this is a vegetable with a high percentage of water contained in the fibres. Trim, cut into 2 in lengths and blanch before free freezing.	3 mins.	Add frozen celery to soups, stews and casseroles or plunge into boiling, salted water and cook until tender to use as a vegetable.	10 mins.
Chicory	4 months.	Must be blanched and used for cooked dishes only as this vegetable contains a high percentage of water. Trim and blanch before freezing and use for cooked dishes.	3 mins.	Cook in boiling salted water with 1 teaspoon sugar added. Serve in a sauce or braise after cooking in a little stock.	8–10 mins.

	Storage time	*Preparation*	*Blanching time*	*Cooking remarks*	*Cooking time*
Corn-on-the-Cob	6 months.	A good freezer provided the kernels are soft and tender. Trim off the outer husk and cut off the stem.	4–6 mins.	Cook straight from the freezer in plenty of fast boiling salted water. Dress with melted butter.	4–8 mins.
Courgettes (Zucchini)	6–9 months.	Pick small, tender courgettes (use large courgettes for ratatouille). Halve lengthwise or cut into thick or thin slices.	1 min.	Cook in boiling salted water until just tender and toss in melted butter. Or allow to thaw and sauté in butter until just tender.	3–4 mins.
Fennel	6 months.	Only suitable to serve cooked. Trim off green tops and any blemished leaves. Cut into slices and blanch in boiling water to which 1 tabs. lemon juice has been added. Free freeze and pack in polythene bags.	3 mins.	The flavour of fennel is quite strong so use it rather sparingly in stews and casseroles or use as a stuffing for fish (no need to thaw if slices have been frozen). Cook in boiling water to serve with another vegetable or to purée.	6–8 mins.
Kohlrabi	12 months.	Use young tender plants not larger than a cricket ball. Leave small golf ball size kohlrabi whole, slice or dice large ones after peeling. Pack in polythene bags (slices can be free frozen).	3 mins. whole 1 min. sliced or diced	A delicious vegetable well worth growing. Put frozen kohlrabi into boiling water, cook until tender and toss in melted butter or serve in a cream sauce.	10–12 mins.

Leeks	2 months unblanched.	Because of their high water content leeks do not freeze very successfully but if you have an excess in your garden you can clean and thinly slice the leeks and freeze them without blanching for up to two months.			
Marrow	12 months. 6 months for halved marrows.	Do not allow marrows to become too large. Remove skin if tough and cut flesh into cubes. Marrows can be halved with the seeds scooped out and frozen for stuffing later. Free freeze diced marrow and pack in polythene bags.	2 mins.	Cook from frozen in a small amount of fast boiling, salted water. Use marrow as a base for a mixed vegetable ratatouille.	10–15 mins.
Mushrooms	1 month unblanched. 3 months in butter. 4 months blanched.	It is only really worth freezing small button mushrooms as the larger field mushrooms go flabby after a time. Wipe small mushrooms and either blanch them or leave them raw. Free freeze and pack in small quantities in polythene bags. Mushrooms can be sautéd in butter before freezing.	2 mins.	Add to soups or stews or use mushrooms in butter as a base for sauces. Add from the frozen state to soups etc. for the last 5–10 mins. of cooking time.	5–10 mins.

	Storage time	*Preparation*	*Blanching time*	*Cooking remarks*	*Cooking time*
Onions	6 months.	Leave small onions whole and blanch. Chop or slice large onions and free freeze (they need not be blanched). Do not freeze whole spring onions. Pack carefully in double polythene bags to prevent odour escaping.	6 mins. whole. 2 mins. sliced.	Very useful for making quickly prepared stews and casseroles. Use the onions straight from the freezer.	
Parsnips	12 months.	Use roots. Peel and cube the roots and blanch in water to which 1 tabs. lemon juice has been added. Freeze and pack in polythene bags.	2–4 mins.	Parsnips are a very underrated veg. Add small quantities of the frozen vegetable to stews and casseroles or serve as a vegetable by cooking the frozen vegetables in fast boiling water for about 10 mins. until tender and tossing in butter. Diced parsnips can also be roasted in oil in the oven.	10–15 mins.
Peas	12 months.	Use young peas and grade if possible. Free freeze and pack in polythene bags.	1 min.	Do not over blanch. Cook peas straight from the freezer in boiling salted water or in melted butter.	4–5 mins.

Peas, Mange Tout or Sugar	12 months.	Top, tail and string if necessary. Blanch, free freeze and pack in polythene bags. Freeze only young tender peas.	1 min.	Do not over blanch. Cook mange tout peas straight from the freezer in fast boiling water. Do not overcook, after about 3–4 mins. they should be tender but still crisp.	3–4 mins.
Peppers, Red and Green	6 months unblanched. 12 months blanched.	Useful additives to stews etc. but cannot be used raw. Peppers can be left whole or have the core and seeds removed and the flesh chopped or cut into thin strips. The flesh can be left raw or blanched for longer storage life. Free freeze chopped or diced peppers and pack in small quantities in polythene bags.	3 mins.	Add frozen chopped peppers to soups, stews and casseroles. Blanch or grill whole peppers and remove the core and seeds for stuffing. Peppers can also be stuffed before freezing.	5–10 mins.
Potatoes, Chipped	12 months.	Peel and chip potatoes and either blanch or part fry until soft but not golden. Drain very well. Free freeze and pack in polythene bags.	1 min.	Add frozen chips to deep, very hot fat. Take care as the fat will spit, and cook until golden brown. Drain off excess fat on kitchen paper (if you are cooking a large quantity re-fry chips just before serving).	

	Storage time	*Preparation*	*Blanching time*	*Cooking remarks*	*Cooking time*
Potatoes, new	6 months.	Very young new potatoes can be frozen if you like to serve these in the winter. Wash and scrape the potatoes and boil them until just tender. Drain well, toss in a generous quantity of melted butter with some finely chopped parsley, cool and pack in rigid containers.		Put potatoes in a casserole. Cover with foil and bake in a hot oven (400°F, 200°C, Reg. 6) for about 40 mins. until hot through. Sprinkle with a little more freshly chopped parsley.	40 mins.
Spinach	12 months.	Trim off tough stalks, wash well to remove all grit, blanch and drain really well getting rid of all possible moisture. Pack in polythene bags.	2 mins.	Plunge frozen spinach into a small quantity of fast boiling water. Drain really well and toss in butter or purée with butter, cream and seasoning. Spinach can also be cooked in melted butter instead of water. Use spinach water to add to vegetable soups.	

omatoes	12 months.	Tomatoes freeze really well in any cooked form. Peel and purée the tomatoes until tender with or without seasoning. Pack in small quantities in rigid containers.		Add tomato purée to soups, stews and casseroles to add flavour or make it into a delicious sauce to serve with any number of savoury dishes. For tomato sauce and purée recipes see page 165. Green tomatoes for tomato chutney can be peeled, chopped and frozen without blanching for short periods.	
Turnips, Swedes and other root vegetables	12 months.	Diced root vegetables can make a useful addition to soups and stews. Cut into small dice for soups and larger chunks for stews. Blanch, free freeze and pack in polythene bags.	3 mins.	Add from the freezer to soups, stews and casseroles or cook in boiling water until tender and use as a vegetable.	10–15 mins.
Salisfy and Scorzonera	6 months.	Salisfy discolours easily so peel the roots dropping them as you do so into well salted water and blanch in water to which lemon juice has been added. Young roots can be frozen whole and larger ones should be cut into sticks.	3 mins.	Cook salsify straight from the freezer in boiling, salted water, drain well and toss in melted butter. Sprinkle with finely chopped parsley before serving.	

Step by step blanching and freezing guide to vegetables

1. Prepare vegetables. Heat water to boiling point. Assemble equipment. Calculate blanching time. (See previous chart.)
2. Place vegetables into blanching basket in small quantities and plunge into boiling water. Time blanching from the moment the water returns to the boil.
3. Remove blanching basket and plunge straight into iced water. Drain as soon as the vegetables have cooled.
4. Spread vegetables on tea towels to dry. Freeze as soon as vegetables have cooled and dried.

Note: If you plan to freeze a large quantity of vegetables it is worth investing in two blanching baskets so that more vegetables can be blanched once one lot is cooling and the water has returned to the boil.

Must vegetables be blanched before being frozen?

When you have a large quantity of vegetables to prepare for the freezer at one time the chore of blanching, cooling and drying them before freezing sometimes seems like that last straw that broke the proverbial camel's back and I have often been told of the excellent results obtained by putting fresh vegetables in the freezer without blanching. In my experience, however, and I have tried and tested every conceivable method of dealing with this problem, the answer for the successful and foolproof freezing of almost all vegetables is to blanch them before freezing. I have found that some vegetables can be frozen without any recognisable deterioration of their quality but, after a month or so, they seem to lose colour and become tough and to have less flavour than they should.

Note: Although experiments are being carried out to try and perfect a method of freezing without blanching these have not yet been successful and I myself would not recommend freezing unblanched vegetables.

Some vegetables you may wish to experiment with by not blanching and by storing for only a short length of time.

Broad beans: Choose only young, tender beans, pod them and free freeze the beans on a tray in the fast freeze section of the freezer until they are solid. Pack the frozen beans in polythene bags, seal, label and return to freezer.

Store the beans for no longer than three months.

Cook the beans straight from the freezer by pouring them into fast boiling, salted water, bring back to the boil and cook until just tender.

French beans: Choose young, thin and tender beans with no strings up the sides. Top and tail the beans and wash and dry them if necessary. Free freeze the beans by spreading them out on a tray and freezing them in the fast freeze section of your freezer and as soon as they are frozen solid, pack them in polythene bags, and seal, label and return to the freezer.

Store the beans for no longer than three months.

Runner beans: Select young tender beans. Top and tail the beans, remove any strings from the sides and slice them thinly. Spread the beans on a shallow tray and free freeze in the fast freezing compartment of your freezer until the beans are solid. Pack in polythene bags, seal, label and return to the freezer.

Store the beans for no longer than three months.

Carrots: Young carrots or diced old carrots can be frozen without blanching for a short length of time although you may find that the carrots have a tendency to turn out rather flabby. Scrape young carrots, pack them in polythene bags, seal, label and freeze.

Scrub, peel and dice old carrots, pack them in polythene bags, seal, label and freeze.

Store for no longer than three months.

Courgettes: If you grow them yourself, courgettes always seem to be ready for picking all at once and you may want to try experimenting with freezing some without blanching them. Wash the courgettes and cut into slices. Free freeze the slices on shallow trays in the fast freeze section of your deep freeze. Pack the slices, as soon as they are solid, in polythene bags and seal, label and return to the freezer.

Store for no longer than one month.

Mushrooms: Again, if you live in the country, mushrooms are available in quantity today but gone tomorrow, so you may want to try freezing a small quantity without blanching, to keep for a few weeks only. Choose small, tightly closed, button mushrooms, avoid washing them if possible, merely wiping off any dirt with a slightly damp cloth and spread out the mushrooms on a shallow tray. Fast freeze the mushrooms until solid, pack in polythene bags, seal, label and return to the freezer.

Store for no longer than one month if you wish to fry the mushrooms – they will tend to toughen after that but can still be used for stews and casseroles.

2. Herbs

Several herbs freeze well and provided they are used in the right way their flavour and appearance are hard to tell from those that are fresh. One word of warning though, herbs do tend to go soggy on being thawed so that sprigs of herbs are not suitable for garnishing. All herbs that require to be finely chopped however, can be frozen stripped from their stalks in small quantities and on removal from the freezer can be crushed between the hands which crumbles the herbs into small pieces. Add the still frozen crumbled herbs to the dish you are making.

If just the flavour of the herbs is required (for stocks and flavouring stews etc.) the stalks can be left on the herbs and they can be frozen in individual 'bouquet garnis' in small polythene bags.

Pick herbs early in the morning on a dry day when the dew has dried but the sun has not yet gained its full strength. Remove the herbs from their stalks or tie into individual 'bouquet garnis'. Package in small quantities in polythene bags, press out as much air as possible. Seal, label and freeze.

Note: In cooked dishes that are frozen for longer than a month the flavour of herbs seems to strengthen and intensify. If you plan to keep pre-cooked dishes with a strong herb seasoning for longer than this time add the herbs on re-heating whenever possible. As this is obviously not possible in dishes such as pies when a lid has been added, cut down on the quantity of the herbs you use when cooking the dish.

Mint sauce: Mint sauce keeps better in the freezer than it will in

11

Preparing and Freezing Fruit

General notes for successful freezing

Most fruits freeze well and great savings can be made by growing or buying and freezing your own fruit, taking advantage of gluts and of seasonable produce. Care must be taken with packaging fruit as it is usually soft and easily bruised or damaged. It is usually preferable to freeze fruit in rigid containers but once frozen solid the block can be turned out and repacked in polythene bags to save using more expensive polythene containers.

Fruit that discolours on preparation must either be treated with lemon juice or ascorbic acid or must be packed in a sugar syrup.

Overripe or bruised fruit can be made into purées or can be prepared for jam making and frozen until you have time to make the jam.

Fruit chosen for freezing should be ripe and in prime condition – fruit will not ripen in the freezer neither will its flavour improve.

Methods of freezing fruit

1. Free flow or open freezing

Soft berry and currant fruits which are easily damaged should be spread on trays and frozen without packaging until they are solid. The frozen fruit can then be packaged and the berries or currants will stay separate during freezing.

Do not wash the fruit unless it is absolutely necessary. If you do have to wash the fruit make sure it is well dried before freezing (use a hair dryer for this). Hull or top and tail the fruit as necessary.

Spread the fruit evenly on shallow plastic trays and open freeze until it is solid throughout. Remove the fruit as soon as it has frozen and pack it immediately into rigid containers (deal with only a small quantity of fruit at a time so there is no risk of the fruit thawing before being packed and returned to the freezer). Do not leave any headspace, seal and label the containers and return the packaged fruit to the freezer.

2. Blanching fruit

Tender fruit which discolours easily like apples benefits from being blanched for a short time before being frozen. This can be done by the steam blanching process which provides less chance of the fruit becoming soggy.

To blanch sliced fruit: Peel, core and slice the fruit and drop the slices as soon as they have been prepared into cold water to which lemon juice has been added or into cold water to which 1 teaspoon of salt to each 2 pints of water has been dissolved. Drain the slices when they have all been prepared, drop them into fast boiling water (see notes on blanching on pages 64–5) and cook for 1 minute after the water has come back to the boil. Remove at once from the water, plunge into iced water, drain and dry the slices on kitchen paper.

To steam blanch fruit slices: A more lengthy process since you cannot usually steam so many slices at one time but it has the advantage that there is no risk of the fruit becoming slightly waterlogged. Place the prepared fruit slices in a steamer over fast boiling water, cover tightly and steam for 2 minutes. Spread the slices on kitchen paper and leave to cool.

3. Dry sugar pack

Fruit that is to be cooked can be frozen with the sugar required to sweeten it. Allow 4 oz of granulated or castor sugar to 1 lb of fruit (use more for damsons or other extremely sour fruit) and use fruit for this method that has a high juice content.

Prepare the fruit for using (i.e. peel, slice etc. as necessary) and place in a bowl. Pour over the required quantity of dry sugar so as to lightly coat the fruit with sugar. Pack in polythene bags or rigid containers allowing a small headspace.

A dry sugar pack is suitable for currants, gooseberries, pineapple, melon, all citrus fruits and rhubarb.

4. Sugar syrup pack

Suitable for all fruits with stones, those that discolour quickly and for easily damaged soft fruit.

In this method fruit is frozen in sugar syrup in which the sugar has dissolved thereby protecting the fruit. Fruit that discolours easily should have either lemon juice or ascorbic acid crystals added to the solution.

The strength of the syrup is determined by the flavour of the fruit and its weight. Sweet, delicately flavoured fruit should be packed in a light syrup whereas a strongly flavoured, sour fruit will benefit from a heavy syrup.

The syrup should be made in advance and allowed to get completely cold before the fruit is added.

Light syrup: 4 oz (100 g) sugar; 1 pint (600 ml) water.

Medium syrup: 11 oz (300 g) sugar; 1 pint (600 ml) water.

Heavy syrup: 1 lb (500 g) sugar; 1 pint (600 ml) water.

For easily discoloured fruit add the juice of 1 lemon to each pint of syrup and either sprinkle the fruit with the lemon juice as it is prepared or add the lemon juice to the cold syrup and drop the fruit into the syrup as soon as it is prepared. If you are using ascorbic acid (this can be bought from a chemist) dissolve ½ teaspoon ascorbic acid crystals in each pint (600 ml) of sugar syrup making sure the crystals are absolutely dissolved before the fruit is added.

Prepare the fruit (i.e. peel, slice etc.), pack it with enough cold sugar syrup to cover into rigid containers leaving ½ inch (1 cm) headspace. Fill the headspace with crumpled foil to prevent the fruit rising above the surface of the syrup, seal, label and freeze.

5. Fruit purées

All purées freeze well and can be made with or without sugar. Cook the fruit with the minimum amount of water required to prevent it from burning and purée when the fruit is tender. Leave the fruit to become absolutely cold before packing and freezing. Leave ½ inch (1 cm) headspace in frozen packs. Add some lemon juice or ascorbic acid crystals to fruit which discolours easily.

6. Stewed fruit
Fruit may be prepared and lightly stewed before being frozen for use when stewed fruit is required.

Thawing fruit
All fruit that is to be served without cooking should be thawed as slowly as possible in its container. Thaw fruit in the bottom of a refrigerator. Fruit that is to be cooked or re-heated can be used straight from the frozen state.

Thawing times for fruit

Soft fruits: 7–8 hours in a refrigerator, 3–4 hours at room temperature.

Note: Allow slightly less time for blackberries, bilberries, strawberries and raspberries which can be served while still slightly frozen.

Hard fruits: Apples, pears, rhubarb etc. 6–7 hours in a refrigerator, 2–3 hours at room temperature.

Fruit purées and fruit packed in syrup: 6–8 hours in a refrigerator 3–4 hours at room temperature.

Fruit mousses and soufflés: Allow to thaw overnight in a refrigerator for at least 12 hours.

Uncooked fruit pies: 2–4 hours at room temperature.

Cooked pies: 2–4 hours at room temperature.

Fruit flans: 6–8 hours in a refrigerator; 3–4 hours at room temperature.

Fruit crumbles: Re-heat from frozen.

Freezing specific fruit

Read the general notes on freezing of fruit at the beginning of this section. The success of fruit freezing depends to a certain degree on the quality of the fruit itself and the conditions under which it was picked. What follows are basic guidelines for the general freezing of fruit but there are, of course, other fruit freezing ideas you may want to try out yourself. So much of home cooking lies in personal experiments.

Apples

Eating apples: Unless you have a supply of eating apples which you have grown yourself and which you are not going to keep in dry store for some reason or which are so battered and bruised you feel it is worth turning them into a purée it is really not worth freezing eating apples as these can perfectly well be stored in a cool, dry, dark place and will only take up valuable room in your freezer.

Cooking apples: Choose any firm, tart cooking apple that is free from bruises, blemishes and discolourations.

Process any blemished or damaged fruit into purée or use it to make apple butters or jellies.

Packing apples dry or in dry sugar packs: Peel, core and slice the fruit dropping the slices into cold water into which 1 tablespoon of salt to every pint of water has been dissolved (this is to prevent the slices discolouring after they have been cut). Rinse in cold water, dry the slices, blanch them in boiling water for 1 minute after the water has returned to boiling point, cool by plunging the slices into iced water and dry them on kitchen paper (or steam blanch them for 2 minutes over boiling water).

Spread the slices on a shallow plastic freezing tray and fast freeze them until they are solid.

Pack the slices with or without sugar in polythene bags or rigid containers, seal, label and return to the freezer.

Use the slices for puddings, pies, for stewed fruit or for fruit purées.

Apples in sugar syrup: First prepare a medium sugar syrup and leave to cool. Prepare and blanch the apples as above. Pack the cooled apple slices in enough sugar syrup to cover. Leave head-space, seal, label and freeze.

Use the slices for stewed apple dishes or for other apple dishes, draining off some of the sugar syrup before using (this can be used again).

Stewed apples: Make in the usual way and cook until the apples are just soft. Leave to cool, pack in rigid containers, seal, label and freeze.

Apple purée: Peel, core and roughly chop the apples and stew them with enough water to prevent them sticking and a little lemon juice until the apples are soft. Purée the apples through a fine sieve, a food mill or in an electric liquidiser or food processor. Use the purée as a basis for apple sauce, as baby food or for recipes which call for apple purée.

Note: Sugar can be added before freezing the apples but on the whole I think it is better to omit the sugar so that the purée can be used either for sweet or savoury dishes as required. Make sure you label the purée 'sweetened' or 'unsweetened'.

Good apple purée can be made from cheap windfall and second quality fruit provided any bruises or blemishes are cut out. It is a useful standby to have in your freezer.

Cooking the frozen fruit: If the slices are to be left whole then it will almost always be worth while thawing your fruit before cooking it or cooking the slices very slowly so they do not break up.

Fruit that is to be stewed or puréed can be thawed over a low heat. Fruit purées and stewed fruit which is to be served hot can be heated straight from the frozen state although this should be done over a low heat and the fruit should be stirred frequently to prevent sticking in the pan.

Avocados

Avocados cannot be frozen without being puréed with lemon juice to prevent discolouration and either sweetened or seasoned as a stabiliser.

Mash or purée avocado flesh with lemon juice, season with salt and pepper or sweeten with sugar and use as a base for dips or for sweet avocado mousses or ice cream.

Pack in rigid containers, seal, label and freeze.

Apricots

Choose plump, firm but strongly sweet smelling fruit that has a good golden yellow colour (you may find it necessary to ripen imported apricots in the home – the best way to do this is to store the apricots in a brown paper bag in a warm, dark and dry place).

Packing apricots in sugar syrup: Do not try to process too many apricots at a time as the fruit will discolour unless the freezing process is carried out reasonably swiftly.

Skin the apricots if their skins are tough by dropping them into boiling water, leaving them for a few minutes and then draining off the water and sliding off the skins.

1. Halve and stone the apricots, cover them with a medium sugar syrup and pack in rigid containers. Leave headspace for expansion, seal, label and freeze.
2. Leave fruit whole, cover with sugar syrup, pack in rigid containers, seal, label and freeze.

Apricot purée: A very useful and well flavoured fruit purée which can be used as a basis for ice cream, mousses, fools and many other dishes.

Halve and stone the fruit and cook with the minimum amount of water with 4 oz (100 g) sugar to each 1 lb (½ kilo) fruit until the apricots are tender. Purée the fruit through a food mill in an electric blender or food processor, cool, pack in rigid containers, seal, label and freeze.

Note: The stones of the apricots can be cracked and the kernels added to the apricot purée to add an overtone of almond flavouring.

Stewed apricots: Halve and stone the apricots (having removed the skins if they are tough) and stew them in ¼ pint (125 ml) water and 4 oz (100 g) sugar to each 1 lb (½ kilo) fruit until the apricots are just tender. Cool, pack in rigid containers, seal, label and freeze.

Cooking the frozen fruit: If fruit is to be eaten without cooking, thaw it overnight in the refrigerator. Fruit that is to be cooked can be re-heated straight from the frozen state or first left to thaw at room temperature.

Bananas

Bananas in their skins, sliced or whole bananas cannot be frozen. The only way you can freeze bananas is to mash or purée them with lemon juice and sugar to use as a sandwich filling. Pack the purée in a rigid container leaving headspace for expansion, seal, label and freeze. Thaw in the refrigerator.

Bilberries (also called blueberries and wortleberries etc.)

Bilberries grow wild in many parts of England but you can also cultivate them in your garden.

Pick when the fruit is dry, preferably in the morning after the dew has dried but before the sun has reached its full heat. Spread the berries on shallow plastic trays and free freeze until the fruit is solid. Pack in polythene bags or in rigid containers, seal, label and freeze.

To cook the frozen fruit: Use the fruit straight from the freezer for puddings and pies.

Blackberries

A prolific 'free' food in the UK. A larger variety can be cultivated in the garden.

Pick when the fruit is dry, preferably in the morning after the dew has dried. Choose large, firm and well coloured berries.

Free freeze blackberries: Spread the berries on shallow plastic trays and freeze until they are solid. Pack the fruit into rigid containers. Seal, label and freeze.

Blackberry purée: Purée overripe soft berries with sugar and lemon juice to taste and rub through a sieve or a food mill to remove the pips. Pack in rigid containers, seal, label and freeze.

To use the frozen berries: Thaw fruit that is to be eaten without cooking in the refrigerator. Cook fruit from the frozen state and use for puddings and pies etc.

Cherries

Red cherries: Dry pack cherries with sugar unless they are very sweet: Choose firm ripe fruit with a good colour. Remove the stalks, wipe the fruit if necessary and free freeze on shallow plastic trays until solid. Pack the fruit in rigid containers with 4 oz (100 g) sugar to each 1 lb ($\frac{1}{2}$ kilo) fruit. Seal, label and freeze.

Black cherries: Pack black cherries in sugar syrup: Make a medium sugar syrup. Stew sour fruit in the syrup until tender, drop sweet fruit into cold sugar syrup. Cool if necessary, pack in rigid containers leaving headspace; seal, label and freeze.

To use the frozen fruit: Thaw fruit that is to be used without cooking in the refrigerator. Thaw fruit that is to be cooked before using.

Currants: black, red, white and yellow

Choose firm ripe fruit and pick in dry weather.

Free freeze currants: Choose large ripe fruit with a good colour, discarding any green berries. Pick over currants, spread them on shallow plastic trays and free freeze until solid. Pack in rigid containers, seal, label and freeze.

Note: Blackcurrants can be frozen on the stalks and the stalks removed while the fruit is still frozen.

Stewed blackcurrants: Stew with the minimum of water and enough sugar to sweeten. Cool, pack in rigid containers, seal, label and freeze.

Currant purée: Stew currants with enough sugar to sweeten and purée (the fruit can be rubbed through a sieve or put through a food mill to remove the pips). Cool, pack in rigid containers, seal, label and freeze.

To use the frozen fruit: Thaw fruit that is to be eaten without cooking in the refrigerator. Fruit that is to be cooked can be used straight from the frozen state.

Damsons

Pick when the fruit is really ripe and wipe if necessary. Pack damsons in heavy sugar syrup with ascorbic acid or stew with enough sugar to sweeten until the fruit is soft. Cool, pack in rigid containers, seal, label and freeze.

To use the frozen fruit: Thaw before cooking or heat gently until thawed and use for puddings and pies.

Gooseberries

Pick when the fruit is in its prime. Top and tail, spread on shallow trays and 'free freeze' fruit until solid. Pack in rigid containers or polythene bags, seal, label and freeze.

To use the frozen fruit: The gooseberries can be stewed from the frozen state and used in the usual way.

Grapefruit

Grapefruit freeze well and are very seasonal, so watch out for prices to drop and buy in bulk.

Freeze grapefruit with or without sugar: Peel fruit, divide into segments and peel off the skin and membrane from each segment. Pack in rigid containers with or without 4 oz (100 g) sugar to each 1 lb ($\frac{1}{2}$ kilo) fruit in between layers of grapefruit segments. Seal, label and freeze.

To use the frozen fruit: Thaw overnight in a refrigerator for the best results.

Grapes

Choose ripe, firm fruit.

Pack grapes in a medium sugar syrup to prevent skins toughening: Make a medium sugar syrup and leave to cool. Leave small seedless grapes whole and strip them from their stalks. Halve large green and black grapes and remove their pips. Pack the grapes in rigid containers, cover with sugar syrup and leave a headspace for expansion. Seal, label and freeze.

Note: If you want to freeze grapes for use in savoury dishes such as *sole véronique*, they can be frozen in a salt and water solution to prevent the grapes being too sweet. Prepare a solution of 1 pint of water with 1 tablespoon salt dissolved in it. Cover the grapes with the solution and freeze as above.

To use the frozen fruit: Thaw grapes before use. Defrost grapes that are to be eaten without cooking in the refrigerator. Grapes that are to be cooked can be thawed at room temperature.

Lemons

Lemons vary considerably during the year and, if, like me, you use them a lot for both sweet and savoury dishes a very worthwhile saving can be made by buying them in bulk when they are at their cheapest. Coarse skinned lemons produce more zest or peel; thin skinned lemons produce more juice.

Freeze lemons whole: Lemons can be frozen whole. Wipe over the skins, pack them individually in small polythene bags, seal, label and freeze.

Sliced lemons: Cut the lemons into thin slices, remove the pips and arrange the slices on shallow plastic trays. Free freeze the slices and pack them in polythene bags when solid.

Grated lemon rind: Lemon rind is an invaluable flavouring ingredient. Grate off the rinds of lemons that are to be used for juice and pack in quantities of ½ a lemon rind at a time in small parcels wrapped in transparent wrapping film. Overwrap in polythene bags, seal, label and freeze.

Lemon juice: Pour lemon juice into small ice cube trays and freeze.

Pack the cubes of lemon juice in polythene bags, seal, label and return to the freezer.

Using frozen lemons: Lemon slices, grated rind and juice can all be used from the frozen state. Whole lemons should have their rind grated off while they are still frozen and then left to thaw at room temperature until suitable for squeezing.

Melons

Melons freeze reasonably well despite their high liquid content. The flesh will become slightly flabby on thawing but is perfectly acceptable for fruit cocktails and salads.

Freezing melons in a sugar syrup: Make a medium sugar syrup and leave to cool. Halve the melon, remove the seeds and scoop out the flesh with a ball scoop or cut it into cubes. Pack the melon in rigid containers, cover with sugar syrup and leave a headspace for expansion. Seal, label and freeze.

Freezing melon to be used for savoury dishes: Melon can be frozen in a dry pack for a short period. Prepare the melon as for a sugar syrup pack in rigid containers. Seal, label and freeze.

To use melon from the frozen state: Thaw melon at the lowest possible temperature, if possible in a refrigerator.

Oranges

Orange segments: Peel oranges, divide into segments and remove all pith, membrane and pips. Free freeze the segments on shallow plastic trays and pack when frozen solid in polythene bags.

Or: Cover the segments with a medium sugar syrup, pack in rigid containers, seal, label and freeze.

Or: Pack layers of segments with 4 oz (100 g) sugar to each 1 lb ($\frac{1}{2}$ kilo) fruit in rigid containers. Seal, label and freeze.

Whole oranges: See lemons.

Sliced oranges, orange peel and juice: See lemons.

Oranges for marmalade: Buy Seville oranges in bulk and freeze whole, packed in polythene bags. See recipe for Marmalade on page 94.

Using the frozen fruit: Thaw frozen orange segments in a refrigerator to use raw. Follow instructions for using lemons for orange slices, peel and juice.

Peaches and nectarines
Choose firm, sweet smelling fruit unmarked by any bruises and blemishes. Peel the fruit by dropping them into boiling water, leaving them for 1 minute and then sliding off the skins. Brush fruit with lemon juice to prevent discolouration.

Freezing peaches and nectarines in a sugar syrup: Make a heavy sugar syrup and add ascorbic acid crystals to prevent discolouration. Halve and remove the stones of the fruit or cut into slices. Cover with cold sugar syrup, pack in rigid containers, seal, label and freeze.

To use the frozen fruit: Thaw in a refrigerator.

Pears
Freezing uncooked pears is not very satisfactory. If you do have a number of hard cooking pears available the best way to freeze them is to peel the pears, remove the core and poach them in a medium sugar syrup with lemon juice added until the pears are tender. Cool, pack in rigid containers, seal, label and freeze.

To use the frozen fruit: Heat gently from the frozen state or leave to thaw in a refrigerator.

Pineapples
Another very seasonal fruit which only comes on the market at certain times of the year. If your family is very fond of pineapples it will probably be worth while buying them in bulk from a wholesaler.

Choose firm fruit with a sweet smell. The fruit should still have some green on the skin or it will be over-ripe.

Dry pack freezing of pineapple: Peel the pineapples, remove the core and cut into rings and slices. Pack in rigid containers and pour over every drop of juice collected as you prepared the fruit. Seal, label and freeze.

Freezing sliced pineapple in sugar: Layer the prepared slices of pineapple in rigid containers with 4 oz (100 g) sugar to each 1 lb ($\frac{1}{2}$ kilo) fruit sprinkled between the layers. Use for fruit salads.

Freezing pineapples in sugar syrup: Make a medium sugar syrup and leave to cool. Cover the prepared pineapple slices with syrup, pack in rigid containers, seal, label and freeze.

To use the frozen fruit: Thaw pineapple in a refrigerator or at room temperature.

Note: Bruised or second quality fruit can be peeled with the core removed and then puréed to use as a flavouring ingredient for ice creams, soufflés and mousses etc. Make sure you cut out all blemishes before puréeing the fruit. The fruit can be frozen with or without sugar.

Plums and greengages
Choose large, firm but ripe fruit.

Pack greengages and other plums in syrup to prevent their skins hardening: Use a heavy syrup for fruit that is on the sour side and a medium syrup for sweeter fruit.

Prepare a suitable syrup with ascorbic acid crystals added and leave to cool. Wipe fruit if necessary, halve, remove stones and pack in rigid containers covered with syrup. Leave a headspace, seal, label and freeze.

Puréed plums: Stewed and sweetened plums with the stones removed can be puréed and used as a base for either sweet or savoury sauces.

Stewed greengages or plums: Stew plums or greengages with sugar and a minimum of water in the normal way. Cool, pack in rigid containers, seal, label and freeze.

Cooking the frozen fruit: Thaw the fruit for using in puddings and pies etc. If the fruit is to be served stewed it can be gently heated from the frozen state.

Raspberries and loganberries
Probably the best fruit for freezing. Properly frozen and thawed raspberries and loganberries are very difficult to tell from fresh fruit that has been left to stand for a time before being served.

Only freeze whole fruit that is in absolutely first class condition. Use second quality fruit for puréeing.

Free freeze raspberries and loganberries without sugar: Sugar draws out the juice of the berries so freeze them as they are for best results. Hull fruit, arrange it on shallow plastic trays and freeze until solid. Pack the frozen fruit in rigid containers, seal, label and freeze.

Freezing raspberry or loganberry purée: Gently cook the fruit with 4 oz (100 g) sugar to each 1 lb ($\frac{1}{2}$ kilo) fruit until tender. Purée through a sieve or food mill to remove the pips. Cool, pack in rigid containers leaving headspace, seal, label and freeze.

To use the frozen fruit: The more slowly you can thaw fruit that is to be served raw the better the results will be. Thaw raspberries in their sealed containers in the refrigerator.

To use the frozen purée: Thaw at room temperature or heat slowly from the frozen state.

Rhubarb

Rhubarb freezes well and is well worth growing for the freezer since the flavour of this fruit makes it suitable for a surprisingly wide variety of dishes. Rhubarb is rather an old fashioned fruit that could be used a lot more right throughout the year.

Choose young, tender stalks.

Rhubarb packed in a dry pack: Wash and trim stalks and cut into 1–2 inch lengths. Free freeze until solid. Pack frozen stalks in rigid containers, seal, label and return to the freezer.

Stewed rhubarb and rhubarb purée: Stew fruit with enough sugar to sweeten and a minimum of water until the fruit is tender. Cool, pack in rigid containers, seal, label and freeze.

To use the frozen fruit: Thaw uncooked fruit at room temperature. Thaw cooked fruit and purées at room temperature or heat gently over a low heat. Use for pies, puddings and as a flavouring for ice cream.

Strawberries

New freezer owners should be warned that strawberries do not emerge from the frozen state in a condition which can be passed off as 'fresh' fruit. They have a high percentage of liquid which forms ice crystals and these break up the fibres of the fruit. As a result frozen strawberries tend to have a slightly flabby texture. Their flavour, on the other hand is unimpaired and if the fruit is cheap or you grow your own, it is worth freezing. I prefer to eat the best of my strawberries fresh and to freeze second quality fruit for use as purée, in fruit salads or as a garnishing material.

Choose firm, ripe fruit (unless you are going to make purée) and hull the fruit before freezing.

Freezing strawberries in a dry pack: Hull the fruit, grade into sizes and free freeze on shallow plastic trays. Pack the solid fruit in rigid containers, seal, label and return to the freezer. Thaw the fruit as slowly as possible in a refrigerator. The fruit can also be sliced before freezing.

Puréeing strawberries for freezing: Purée the fruit in an electric liquidiser or through a food mill. You can sweeten the purée or leave it unsweetened. Pack the purée in rigid containers, seal, label and freeze.

Using the frozen fruit: Thaw strawberries that are to be eaten whole or in slices at the lowest possible temperature, preferably in a refrigerator. Strawberry purée can be thawed at room temperature or if the purée is to be cooked, over a low heat.

Nuts

Nuts keep better in the freezer than they do in a larder. Freeze nuts in the shell, shelled, blanched or chopped. Roast nuts can also be frozen. Pack in small quantities in polythene bags and thaw at room temperature.

Fruit salads and cocktails

Fruit salads and cocktails can be frozen very successfully. Avoid fruit which will discolour and cover the mixed fruits with a medium sugar syrup. Pack in rigid containers, seal, label and freeze.

Note: Do not add bananas, apples or pears to fruit salads until just before serving.

To use frozen fruit salads or cocktails: Thaw at the lowest possible temperature as slowly as possible. For the best results thaw in a refrigerator overnight, for serving the next day.

Freezing fruits for jam making

There are times when you may have fruit from your garden or may have access to inexpensive, second class fruit for freezing but when you may not have the time to actually process the fruit for jam making. The fruit can be frozen when it is available and then the jam made later.

Jam from frozen fruit is made in the same way as it would be with fresh fruit, using the same quantities, but, since a little of the pectin may be destroyed during the freezing, you may possibly have to add a little extra sugar to get a good set.

Prepare the fruit for jam making, pack it with or without sugar, making a careful note on the package of the sugar content and stew the fruit from the frozen state.

Note: With some fruit with taut skins, like currants, there may be a slight toughening of the skins during the freezing process and it is worth stewing the fruit lightly before freezing.

Crystallised and dried fruit
As these products keep well in a dry larder there is not much point in freezing these fruits unless you can buy them exceptionally cheaply and want to keep them for a considerable length of time. They freeze well and can be used straight from the freezer for most purposes.

Candied peel
Candied peel can be frozen but there is not very much point in doing this unless you prefer to make your own in which case it might not have as long a storage life as a commercially candied product. Pack candied peel for the freezer in small quantities. It can be used staight from the freezer.

Mincemeat
Homemade mincemeat can also be frozen.

Seville marmalade
Making marmalade is a messy business so resign yourself to it. I cut my oranges up on a plastic tray so that the juice does not run all over the place.

If you are using frozen oranges the fruit should be cooked whole before cutting and can be cooked straight from the frozen state.

3 lb Seville oranges
5 pints water
6 lb preserving sugar
juice and finely grated rind of
  2 lemons

Put the oranges into a large preserving pan, cover with 4 pints of water, bring to the boil, cover and simmer gently for 1½ hours or until the skins of the oranges are really soft. Lift the oranges out of the water with a slotted spoon, cut them in half, scoop out the pips and the white pith from the centre of the oranges with a small spoon and put the pith and pips into a small saucepan with 1 pint water. Bring to the boil and simmer for 15 minutes to extract all the pectin from the pips.

Cut the oranges into very thin slices. Return the oranges to the water in which they were cooked. Add the strained liquid from the pith and pips, add the sugar, lemon juice and rind, and stir over a low heat until the sugar has all dissolved. Bring to the boil and cook over the highest possible heat so that the mixture is boiling very fast for 20 minutes. Test for a set by placing a little of the marmalade on a saucer (a skin should form over the surface if the marmalade is ready). Take the pan from the heat, skim off any scum from the surface and spoon into warm, clean jars. Cover with waxed paper discs while still warm but leave to seal until completely cold.

PART TWO: RECIPES

Introduction

Cooking ahead for freezer storage

The deep freeze, if properly used, can be a real boon for the busy housewife, hostess or working mother since not only fresh produce but also pre-cooked foods can be stored for certain periods so that food can be cooked when it is convenient.

There are obviously some pre-cooked foods which freeze more successfully than others (amongst these are pies, stews, casseroles, cakes, pâtés and mousses) and I think it is important to point out that pre-cooked food should never be kept too long in the freezer or it will lose flavour and sometimes deteriorate in texture. Try to think of your freezer in terms of short rather than long term storage and keep a careful record of the contents so that pre-cooked dishes do not get overlooked.

Most dishes will adapt well to freezing provided you follow a basic set of guidelines.

1. Do not freeze dishes that contain hard boiled eggs or cooked, whole or chopped potatoes.
2. Spices and flavourings tend to intensify if a dish is left for any length of time in the freezer so it is as well to under rather than over flavour before freezing. Always test for seasoning and flavouring before serving a frozen dish.
3. Do not freeze dishes made with cream or egg yolks as these will separate. The cream and eggs can be added when the dish is thawed.
4. Make sure that meat is well covered by a sauce in its container so that it is protected from drying out.
5. Do not overcook pre-cooked foods for the freezer. Remember

that there will be some extra cooking time when you re-heat the dish and it is better to allow a little longer for this process rather than to overcook the ingredients before freezing.

6. Prepare your favourite dishes in the normal way taking note of the comments above. Cool them as quickly as possible, pack, seal, label (giving thawing and re-heating instructions), and freeze. They should be completely cold before being packed and frozen and care must be taken to ensure that they are thoroughly re-heated before being served – nothing is more off-putting than coming across a lump of ice in the middle of what is meant to be a hot dish.
7. Try not to freeze pre-cooked dishes for much longer than two months.

'Three step' dishes for the freezer

Monotony is one of the great killers to the enjoyment of food and although it may be convenient to make fifteen cottage pies at one time you will find both you and your family get very bored of eating exactly the same food over and over again. In order to prevent this monotony and to provide greater scope in your cooking you can, instead of freezing finished dishes, prepare, cook and freeze basic recipes which can then, on being thawed, be used as the basis for one of many possible dishes. I find it invaluable to keep a store of basic minced meat, beef, lamb and chicken stews or casseroles in the freezer. When I require something in a hurry I then have, at my fingertips, a coq au vin, or chicken with paprika. The basic hard work that goes into this kind of dish is all done in advance and all you have to do before serving is to alter the taste to suit the occasion or your mood.

In the 'three step' method for freezing pre-cooked basic recipes the first step is the cooking and freezing of the dish; the second step is the thawing of the frozen dish and the third step involves the addition of extra ingredients to transform it into the finished article. Ideas for variations on the different basic recipes are given in the following pages but in many cases it will be your own imagination which comes to the fore here as well as, of course, the ingredients you have at hand.

The great advantage of this form of pre-cooking and re-heating means that not only can you prepare the dishes when you have the time and inclination to spend some time in the kitchen but the finished dish has a freshness and individuality which you never get from cooking and freezing large quantities of the same recipe.

Suitable dishes for 'three step' freezing
Basic consommé and vegetable soups, minced meat recipe, basic meat ball recipe, stews, casseroles and hot pots, pancakes, cooked and uncooked flan cases, vegetable stews, hamburger mix, fish stew, fruit salad and stewed fruits.

Each recipe section which follows contains suggested three-step recipes, which are marked with an asterisk. Do be imaginative when you re-heat and finish the dishes.

Soups

There are so many soups that freeze well that unfortunately one cannot begin to list them in one general book on freezing. I have included some of my favourites and by following the basic guidelines for freezing soups you can use your imagination and adapt almost all your favourite soups for freezing.

Do's and don'ts on freezing soups

Don't add potatoes or other floury vegetables to a soup before freezing, unless the root vegetables are purėed before being added to the soup.

In principle, don't thicken soups before freezing. Thickenings can easily be added when you are thawing the soups and if they are added before freezing the soups have a tendency to curdle.

Don't add cream or egg yolks to soups before freezing for the same reason – they will curdle on re-heating.

Keep seasonings light. You can always add more seasoning when you re-heat.

Don't incorporate garlic into your soups before freezing as it will tend to develop an 'off' taste – again this can be added when re-heating the soup.

Don't add garnishing before freezing. In order to give the desired 'fresh' effect, garnishings should be added just before serving.

Always remove every particle of fat from your soups before freezing. Fat tends to develop a rancid flavour in the freezer and may ruin the flavour of your otherwise successful soups.

Do leave at least ½ inch headspace when freezing soups as they will expand in the freezer.

Stocks for freezing

In our grandparents' days making home-made stock was as much an everyday event as watching the news on television is now. For most of us it is not practical to have a huge, regenerating stock pot on the boil day in and day out but, on the other hand, with a deep freeze in your home, it makes good sense to make your own stock (especially if you have a pressure cooker) when you have the time and to freeze the results making a useful bank of well flavoured soup, stew and casserole bases at hand when they are needed. Stock also makes an excellent base for a good many exciting sauces and is probably one of the most useful step by step items to keep in your freezer larder.

Stock cubes are useful and can be added to stock for extra flavour but they do not, and can never, imitate the individual flavour, strength and goodness of a home-made stock.

Increase the strength of stocks by straining and then boiling them. Store double or treble stock in frozen cubes and reconstitute by adding water when re-heating.

Store unreduced stocks in polythene containers or freeze in rectangular containers until solid, turn out and pack in polythene bags for long term storage.

Basic meat stock

(Makes 4 pints)

This is a well flavoured stock that can be used as a base for consommé or for any meat or vegetable soups that do not require a chicken or 'white stock'.

1 lb shin beef
2 lb beef bones
(preferably marrow bones)
4 pints water
2 carrots
1 onion
6 peppercorns
½ teaspoon salt
1 stick celery
4 sprigs parsley
2 bay leaves

Place the bones in a roasting pan and roast them in a hot oven (400 F, 200 C, Reg. 6) for about 15 minutes until lightly browned. Drain off and reserve the fat in the pan. Place the bones in a heavy bottomed saucepan, pour over the water, add the shin beef, cut into small pieces, the peppercorns and salt bring to the boil, lower the heat, cover and simmer gently for two hours.

Scrub and chop the carrot. Wash the onion but leave on the outer skin. Roughly chop the onion and skin. Roughly chop the celery. Heat the fat from the bones in a saucepan. Add the vegetables to the fat and cook them over a medium heat, stirring to prevent sticking, until the onion is coloured and the fat has been absorbed. Add the vegetables to the stock with the parsley and bay leaves. Return to the boil, lower the heat, cover and simmer for a further hour.

Strain the stock through a sieve lined with muslin into a large bowl and leave to cool.

To freeze: Refrigerate until the fat has formed a firm skin over the surface. Remove the skin of fat and pack in polythene containers or freeze in a rigid container until solid and then turn out and repack in a polythene bag. Seal, label and freeze.

To use: Turn the frozen stock into a saucepan and heat over a low heat until thawed.

Note: Extra colour and flavour can be given to the stock by adding some over-ripe tomatoes, mushroom peelings and some extra onion skins.

To reduce the stock and double its flavour, turn the strained stock into a heavy bottomed saucepan after removing the fat and bring it to the boil. Boil the stock over a high heat until it is reduced to half. Cool, pour into ice-making cubed trays and freeze until solid. Turn out the frozen cubes, wrap each one quickly in plastic film and pack in polythene bags. Seal, label and freeze.

Use the beef from the stock as a basis for made up meat dishes like 'shepherd's pie'. Or cut it into thin strips, marinade it in a French vinaigrette with freshly chopped herbs, and chill for at least three hours before serving as a summer dish with salad.

Basic chicken stock

1 chicken carcass and giblets
3 pints water
1 chicken stock cube
2 carrots
1 large onion
2 sticks celery
2 sprigs parsley
2 bay leaves
pinch salt
8 white peppercorns

Break up the chicken carcass and put it into a large, heavy bottomed saucepan with the herbs. Wash but do not peel the onion and cut it into quarters. Peel and roughly chop the carrot and roughly chop the sticks of celery leaving the leaves on. Add the vegetables, stock cube, salt and pepper to the chicken and cover with cold water. Bring to the boil, cover tightly and simmer for 2 hours.

Strain the stock into a bowl, through a sieve lined with muslin.

To freeze: Leave to cool and then refrigerate until the fat has formed a skin over the surface. Skim off the skin, transfer the stock to rigid containers and freeze until solid. Turn out, repack in polythene bags, seal, label and freeze.

To use: Turn the frozen stock into a saucepan and heat over a moderate heat until hot through. Use as the basis for soups or for making chicken consommé.

Stock from poultry or a game carcass

1 cooked chicken, duck or turkey etc. carcass
1 large onion
3 carrots
2 sticks celery
1 strip lemon peel
2 sprigs parsley
2 bay leaves
salt and freshly ground black pepper

Break up the carcass and put it into a large heavy saucepan. Wash the onion, leave the skin on and cut the onion into quarters. Scrub and roughly chop the carrots. Roughly chop the celery leaving the leaves on. Add the vegetables to the carcass with the lemon peel and herbs. Season with salt and pepper.

Cover completely with cold water, bring slowly to the boil, skimming off any scum that rises to the surface, cover tightly and simmer over a low heat for two hours. Strain the stock into a clean saucepan through a sieve lined with muslin. Re-heat and check flavour – if the stock has not enough strength, add a stock cube to boost the taste.

To freeze: Cool, refrigerate until the fat forms a skin over the surface, remove the skin and pour the stock into rigid containers. Freeze until solid, turn, out, repack in polythene bags, seal, label and freeze.

To use: Turn the frozen stock into a saucepan and re-heat over a medium heat until hot through. Use as a basis for soups, gravy and sauces or to add to stews and casseroles.

To clarify stocks for consommés and aspics

A good stock of home made consommés in your freezer will provide you with any number of delicious soup bases for both hot and cold beverages and the consommé can also, with the addition of a little gelatine, make a perfect aspic for cold and summer dishes.

3 pints basic meat or chicken stock
1 egg white
1 egg shell

Put the stock into a clean saucepan, bring to the boil and boil over a high heat until the liquid is reduced by one-third.

Beat egg white until just stiffening. Break the egg shell into small pieces.

Add the egg shell and egg white to the stock and bring slowly back to the boil, lightly whisking the egg whites and shell into the stock using a wire or rotary whisk. Boil for three minutes without whisking, remove from the heat and leave until the froth subsides. Return the saucepan to the heat, bring back to the boil once more and then leave to settle again.

Strain the soup through a double thickness of muslin. By this time the stock should be absolutely clear and a rich colour. If it is not clear, repeat the clarifying process with a second egg white and shell.

To freeze: Cool the consommé, pour into rigid containers and freeze until solid. Turn out and repack in polythene bags, seal, label and freeze.

To use: *For hot consommés:*
Turn the frozen consommé into a saucepan and heat over a medium heat until hot through. Add the juice of ½ lemon and 2/4 tablespoons sherry or Madeira and garnish with any of the following: Thin slices of lemon or orange; some grated raw carrot; very thin strips of pancake; finely chopped chives or celery tops; a little well drained tuna fish or truffle peelings.

Note: You can incorporate a tablespoon of tomato purée to give added flavour to the consommé.

For cold consommés: Thaw the consommé in a refrigerator for about 8 hours or overnight. Break up the chilled, jellied liquid with a fork, add the juice of ½ lemon and a little sherry, and serve it in chilled bowls with a garnishing of any of the following: Thin slices of orange or lemon; very finely chopped, peeled tomatoes, with the core and seeds removed; finely chopped chives; a dollop of sour cream topped by some red caviar, finely chopped red or green peppers or some grated cucumber.

Vegetable stock

When you are blanching a large quantity of vegetables at a time (unless they have a very strong flavour) you can use the same water for blanching again and again. During the blanching process the water will pick up a certain amount of flavour from the vegetables and this water should be used for stocks or for making a stock from vegetables only. The stock is particularly useful for adding to a vegetable soup in the place of plain water adding both flavour and body to the soups.

3 carrots	3 pints vegetable water (or water and 1–2 stock cubes)
2 large onions	salt and pepper
2 tomatoes	2 bay leaves
2 sticks celery	3 sprigs parsley
1 small turnip	

Wash and trim but do not peel the carrots and onion and chop them roughly. Quarter the tomatoes. Roughly chop the celery leaving the leaves on.

Peel and roughly chop the turnip. Combine the vegetables in a large, heavy saucepan. Add the herbs, season with salt and pepper. Cover with the vegetable liquid (or the water and stock cubes). Bring to the boil, cover tightly and simmer for 1½ hours.

Strain the stock through a fine sieve and leave to cool.

To freeze: Pour the stock into rigid containers and freeze until solid. Turn out and repack in polythene bags. Seal, label and freeze.

Note: For a little extra flavour you can sauté the vegetables gently in a little butter before adding to the liquid but in that case all the butter must be completely absorbed before the liquid is added.

To use: Turn the frozen stock into a saucepan and heat over a medium heat until hot through. Use for soups, stews and casseroles.

Fish stock

When you buy filleted fish from your fishmonger make sure he gives you the heads, tails and trimmings of the fish as well. These trimmings contain invaluable flavouring material as well as a considerable amount of goodness and vitamin content and they can be utilised to make a well flavoured fish stock which can be used to cook the fish in or to add flavour to the sauce you plan to serve with your fish.

The stock can be boiled after straining to reduce it and intensify the flavour and it keeps well in the freezer.

2 lb fish bones, heads and trimmings
1 onion
2 sprigs parsley
pinch thyme
2 bay leaves
three slices lemon
½ pint white wine
1½ pints water
salt and 12 white peppercorns

Peel and chop the onion. Combine all the ingredients in a large saucepan and bring slowly to the boil skimming off any scum that rises to the surface. Cover and simmer gently for 1 hour. Strain through a sieve lined with muslin and leave to cool.

To freeze: Turn into rigid containers and freeze until solid. Repack in polythene bags, seal, label and freeze.

To use: Turn the frozen stock into a saucepan and heat slowly until the stock is thawed through.

Soup Mareshka (*Serves 5–6*)

1½ pints basic chicken stock
1½ oz butter
1 large onion
2 oz long grain rice
2 bay leaves
2 teaspoons lemon juice
salt and freshly ground black pepper
¼ teaspoon finely chopped tarragon
½ pint milk
1 tablespoon flour

Peel and finely chop the onion. Heat the butter in a saucepan, add the onion and cook over a low heat, stirring to prevent sticking until the onion is soft and transparent. Add the rice and stir for 30 seconds until the rice is transparent. Add the bay leaves, lemon juice and thawed chicken stock. Season with salt and pepper, bring to the boil, stir well, and simmer for thirty minutes. Mix the flour to a smooth paste with a little of the milk. Whisk in the remaining milk using a wire whisk and add all the mixture to the soup with the tarragon. Bring to the boil, stirring all the time until the soup has thickened. Simmer for two minutes, remove the bay leaves, check seasoning and serve piping hot.

Chicken satin soup

(Serves 4)

2 pints frozen basic chicken stock
¼ pint double cream
2 egg yolks
4 oz cooked chicken
1 tablespoon finely chopped parsley

Finely chop the chicken, add it to the stock and simmer for 3 minutes. Heat the chicken stock to boiling point, then lower the heat to just below simmering.

Beat the egg yolks with the cream until really smooth. Add two tablespoons of the chicken soup to the egg and cream mixture and beat well. Add this mixture to the stock stirring vigorously as you pour it in to prevent the soup curdling. Take care the soup does not boil. Continue to stir until the soup is hot and satiny, mix in the parsley, check the seasoning and serve.

Iced chicken and tomato soup

(Serves 4)

1 pint basic chicken stock
1 sliced onion
1 small, inside stick celery
2 green peppers
½ pint tinned tomato juice
1 small bunch watercress

Peel and very finely chop the onion. Very finely chop the celery and celery leaves. Halve the peppers, remove the seeds and very finely chop the flesh.

Combine the vegetables in a saucepan, add the thawed stock and bring to the boil. Cover and simmer gently for 20 minutes. Check seasoning.

Turn into a bowl, cool and refrigerate until well chilled.

Add the tomato juice to the chilled soup, check the seasoning again. Remove the stalks and very finely chop the watercress. Add the chopped watercress to the soup and serve well chilled.

Lettuce soup

(Serves 6)

An ideal recipe for those who grow their own lettuces and find they have more than they know what to do with at certain times of the year. Milk, egg yolks and cream are added to the soup when it is re-heated.

12 oz lettuce leaves
1 small onion
1½ oz butter
1 pint chicken stock
salt and freshly ground black pepper
pinch sugar
pinch ground nutmeg

Wash, dry, and roughly chop the lettuce leaves. Peel and roughly chop the onion. Melt the butter, add the onion and cook over a low heat until the onion is soft and transparent. Add the lettuce and continue to cook over a low heat until the butter has been absorbed and the lettuce has softened. Add the stock, season with salt, pepper and a little sugar and nutmeg. Bring to the boil, cover and simmer for 15 minutes. Drain the soup and purée the lettuce and onion through a fine food mill, a fine sieve, in an electric liquidiser or a food processor. Add the liquid to the purée.

To freeze: Cool the soup base, turn into rigid containers, seal, label and freeze.

To use:
¾ pint milk
1 large egg yolk
4 tablespoons single cream
2 tablespoons finely chopped watercress or celery leaves

Turn the soup into a heavy saucepan. Thaw over a low heat until hot through. Add the milk, bring to the boil and then lower the heat. Beat the egg yolk with the cream until smooth. Add the egg yolk and cream mixture to the soup and heat through, stirring, without boiling, until the soup has thickened and is velvety. Check seasoning and garnish with finely chopped watercress or celery leaves.

Tomato soup*

Step 1: Basic tomato soup (*Serves 8*)

2 lb ripe tomatoes or tinned tomatoes
1 carrot
1 onion
2 bay leaves
½ teaspoon grated lemon rind
6 peppercorns
pinch oregano

1 tablespoon olive or vegetable oil
salt, pinch sugar
$2\frac{1}{2}$ pints chicken stock

Roughly chop the tomatoes. Peel and chop the onion. Peel and chop the carrot.

Heat oil in a saucepan, add the onion and carrot and cook over a low heat until the onion is soft and most of the oil has been absorbed into the vegetables. Add the tomatoes, bay leaves, lemon rind, peppercorns, oregano and sugar, season with a little salt and mix in the stock. Bring to the boil, cover and simmer for 45/60 minutes with the lid half on and half off the saucepan.

Rub the soup through a sieve (no good using a food mill or liquidiser for this as you have to get rid of the pips). Add some more stock if the soup is too thick.

Step 2

To freeze: Cool the sieved soup, pack in 2 rigid containers and freeze until solid. Repack in polythene bags, seal, label and freeze.

Step 3

To use: Turn the frozen tomato soup into a saucepan and heat over a low heat, stirring now and then until the soup is melted.

Cream of tomato soup (*Serves 4*)

Step 3 ingredients

$\frac{1}{2}$ quantity of basic tomato soup
$\frac{1}{4}$ pint single cream

Heat the basic tomato soup until boiling and reduce the heat to below simmering. Add the cream, stir it in with a wire whisk making sure the soup does not boil, check seasoning and serve with *croûtons* of crisply fried bread.

Potage de tomates avec l'orange (*Serves 4*)

Step 3 ingredients

$\frac{1}{2}$ quantity of basic tomato soup
juice of 1 orange and very thinly pared rind of $\frac{1}{2}$ orange
$\frac{1}{4}$ pint single cream

Add the orange juice to the thawed soup and heat until boiling. Lower the heat to below simmering, add the cream and mix it in with a wire whisk.

Cut the thinly pared orange rind (all the white pith should be removed) into very thin stips. Pour the soup into bowls and add the orange rind.

Tomato soup with carrots and curry flavouring (*Serves 6*)
Step 3 ingredients
A delicious rich and satisfying soup with a slight curry flavouring.

½ quantity basic tomato soup
2 carrots
½ pint chicken stock
1 small onion
1 tablespoon olive or vegetable oil
1 teaspoon curry powder
1 teaspoon lemon juice
4 tablespoons sour cream
1 tablespoon chives

Turn the soup into a saucepan and heat gently until thawed through. Peel and finely chop the onion. Peel and chop the carrots. Heat the oil in a large saucepan. Add the onion and carrot and heat over a low heat, stirring to prevent sticking until the onion is soft and the oil absorbed. Mix in the curry powder and cook, stirring for 2 minutes.

Add the stock and lemon juice, cover and cook for 20 minutes until the carrot is soft.

Purée the carrot and onion through a fine sieve, a food mill or in an electric liquidiser or food processor. Add the carrot purée to the tomato soup, mix well, heat through and adjust seasoning. Pour into heated serving bowls and serve with a spoonful of sour cream topped with a sprinkling of chives in each bowl.

Spiced tomato soup (*Serves 6*)
This is a really delicious recipe which is different.

Step 3 ingredients
½ quantity of basic tomato soup
½ pint chicken stock
1 teaspoon paprika pepper
1 glass red wine
1 teaspoon lemon juice
¼ pint double cream
1 small green pepper

Add the paprika, chicken stock and red wine to the thawed tomato soup, bring to the boil and simmer for 15 minutes. Whip the cream until just thick and blend in the lemon juice. Season the cream lightly with salt and pepper.

Cut six thin rings from the green pepper and remove the core and seeds from the rings. Pour the soup into heated serving bowls, float the pepper rings on the surface of each bowl and put a large spoonful of the flavoured cream in the middle of each ring. Sprinkle over a little extra paprika pepper and serve at once.

Brussels sprouts purée* (*Serves* 6)

Make this when you have a glut of sprouts or when they are larger than they should be for serving as a vegetable.

Step 1

1½ pints well flavoured stock
2 lb Brussels sprouts
½ lb potatoes
1 onion
2 oz butter

Peel and chop the potatoes. Clean and trim the Brussels sprouts Peel and chop the onion.

Melt the butter in a saucepan. Add the onion and potatoes and cook over a low heat, stirring every now and then until the butter is absorbed. Add the sprouts and toss over a medium heat for two minutes. Add the stock, season with salt, pepper and pinch nutmeg, bring to the boil, cover tightly and cook for about 20 minutes until the sprouts are tender. Purée the mixture through a sieve, a food mill or in an electric liquidiser or food processor.

Step 2

To freeze: Leave the purée to cool and then pack in a rigid container. Seal, label and freeze.

Step 3 Brussels sprouts soup

½ pint milk
¼ pint single cream
2 oz flaked almonds

Thaw the purée overnight or thaw over a low heat, stirring as the purée melts. Add the milk, bring to the boil and adjust the seasoning. Lower the heat and stir in the cream. Add a little extra milk or stock if the soup is too thick.

Roast the almonds in a hot oven for 2 minutes until they are crisp and golden and sprinkle them over the soup just before serving.

Favourite carrot and orange soup *(Serves 4)*

This is such a freezer favourite that I seem to come across it again and again and it is always a winner. I serve it hot in the winter and chilled in the summer and vary the garnishings to make it just a little different from everyone else's version.

Make the soup when carrots are at their cheapest in November or December or at the end of the season when the winter stock is being sold cheaply.

1 lb 4 oz carrots
1 small onion
1 leek
the grated rind and juice of 2 oranges
2 pints chicken stock
½ teaspoon curry powder
salt and freshly ground black pepper
1 teaspoon brown sugar
2½ tablespoons olive or vegetable oil

To finish: 1 clove garlic.
For garnishing: 4 tablespoons sour cream; 2 tablespoons finely chopped chives or spring onion tops.

Peel and chop the carrots and onion. Remove the green leaves from the leek and chop the white centre. Heat the oil in a saucepan. Add the vegetables and cook over a low heat, stirring every now and then, covering the pan and cooking for 10 minutes until the vegetables are soft and the oil has been absorbed. Add the orange rind and juice, the stock, curry powder, brown sugar, and season with salt and pepper. Bring to the boil, cover and simmer until the carrots are tender.

Purée the soup through a food mill or in an electric liquidiser or food processor.

To freeze: Turn into rigid containers, seal, label and freeze.

To use: Transfer the frozen soup to a saucepan and heat through over a low heat until the soup is piping hot. Serve with a spoonful of sour cream floating on the top of each serving topped with a sprinkling of chopped chives.

If the soup is to be served cold, leave it to thaw in the refrigerator for about 8 hours, beat it with a wire whisk, adjust the seasoning and float the cream on the top as above.

Meat and Game

Almost all forms of meat and game respond well to being deep frozen and meat, particularly, if bought in bulk and at certain times of the year, can provide a great saving for the housewife. Try to establish that meat and game which is to be frozen has been properly hung in order to tenderise and flavour the flesh and, whenever possible, prepare joints, cuts of meat and birds etc. for cooking before freezing them.

Marinading meat in the freezer

Cheaper cuts of meat benefit enormously from being marinaded in the freezer. The longer the meat can be left in the freezer in its marinade the better it will be so when you have access to some cheap cuts of meat, venison, game, or second quality chicken joints, make a marinade, prepare and pack the meat in polythene containers, pour over the marinade (ingredients should be covered with the liquid) and seal, label and freeze.

Store marinaded meats or chicken joints for up to 3 months

Basic marinade for venison and other game

½ pint red wine
½ pint cider vinegar
1 onion
bouquet garni
salt and freshly ground black pepper

Peel and slice the onion. Combine all the ingredients and pour them over the venison or game.

Basic marinade for beef

½ pint red wine
¼ pint vegetable oil
2 tablespoons cider vinegar
1 onion
2 carrots
3 bay leaves
pinch allspice
salt and freshly ground black pepper

Peel and slice the onion. Peel the carrots and cut them into thin sticks. Combine all the ingredients, season with salt and pepper and pour the marinade over the meat.

Basic marinade for lamb or chicken

½ pint red or white wine
¼ pint vegetable oil
juice of 1 lemon
1 piece lemon peel
1 onion
sprig rosemary
2 bay leaves
salt and freshly ground black pepper

Peel and slice the onion. Combine all the ingredients and pour them over the lamb or chicken which should be cut into cubes or, in the case of chicken, joints.

Barbecue marinade

¼ pint red wine
3 tablespoons tomato ketchup
juice of 1 lemon
1 tablespoon soft brown sugar
1 teaspoon made English mustard
few drops Worcestershire sauce
few drops Tabasco sauce
salt and freshly ground black pepper
¼ pint vegetable oil

Combine all the ingredients and mix well. Pour the marinade over the ingredients (there may not be enough marinade in this case to ensure the ingredients are fully covered so prick the joints with a fork and brush them with the marinade before freezing).

To use: Thaw the meat or joints overnight in the refrigerator. Drain off the marinade and follow your favourite recipe for stews, grilled kebabs, casseroles or barbecued meats and poultry etc. Use the marinade as part of your cooking liquid or for basting the ingredients as they cook.

Lamb casserole★ (*Serves 4*)

Make four or more quantities of the basic stew at one time.

Step 1

1½ lb lean lamb
1 large onion
1 oz lard
8 oz carrots
salt and freshly ground black pepper
¾ pint stock
bouquet garni
2 bay leaves

Cut the lamb into ½-inch cubes. Peel and chop the onion. Clean and chop the carrots.

Heat the lard in a frying pan. Add the meat and brown over a high heat, turning the meat as it cooks. Transfer the lamb with a slotted spoon to a casserole and add the onions and carrots to the juices in the pan. Cook over a low heat until the onion is soft and transparent and the fat has been absorbed into the vegetables.

Add the onions, carrots, the bouquet and bay leaves to the meat, pour over the stock, season with salt and pepper and cover tightly.

Cook in a moderate oven (350°F, 180°C, Reg. 4) for 1½ hours or until the meat is tender.

Step 2
To freeze: Remove the bouquet garni, and the bay leaves. Cool the casserole, turn into a rigid container, seal, label and freeze.

Step 3
To use: Thaw the casserole overnight and continue as in the following recipes.

Curried lamb casserole (*Serves 4*)
Step 3 ingredients
1 quantity thawed basic lamb casserole
2 cloves garlic
1 piece fresh root ginger, about ½ inch square
1 tablespoon vegetable oil
1 tablespoon curry powder
1 teaspoon turmeric
1 cooking apple
1 oz sultanas

Peel and crush the garlic. Finely grate the ginger root; heat the oil in a large, heavy saucepan, add the garlic and ginger and cook over a low heat, stirring all the time until the garlic is soft and transparent. Add the curry powder and turmeric and fry over a low heat, stirring, for two minutes. Add the thawed lamb casserole, bring slowly to the boil and simmer for 30 minutes. Peel and chop the apple.

Add the apple and sultanas to the casserole, mix well, check seasoning and continue to simmer for a further 15 minutes. Serve the curry with rice and curry accompaniments.

Casserole of lamb with peas and mint (*Serves 4*)
Step 3 ingredients
1 quantity thawed basic lamb casserole
6 oz frozen peas
1 tablespoon finely chopped fresh mint
1 carton (5 fl oz) sour cream

Turn the thawed casserole into a large, heavy saucepan. Heat through until boiling. Add the peas and simmer for about 15 minutes or until the peas are tender. Check seasoning, stir in the mint and sour cream and heat through without boiling.

Mediterranean lamb casserole (*Serves 4*)
Step 3 ingredients
1 recipe thawed basic lamb casserole
1 small tin tomatoes (7 oz)
1 tablespoon tomato purée

pinch oregano and thyme
1 teaspoon cumin powder
1 green pepper
2 cloves garlic
1 tablespoon olive oil

Roughly chop the tomatoes. Remove the core and seeds of the pepper and cut the flesh into thin strips. Peel and crush the garlic. Heat the oil in a large, heavy saucepan. Add the garlic and pepper and cook over a low heat, stirring, until the garlic is soft and transparent. Add the thawed casserole, the tomatoes, tomato juice and purée and mix in the herbs and cumin. Bring to the boil, cover tightly and simmer gently for 30 minutes. Check seasoning before serving.

Lancashire hot pot*

(Serves 4)

Step 1

1 lean joint best end of neck divided into cutlets
½ pint chicken stock
1½ tablespoons tomato purée
2 lambs' kidneys
2 large onions
4 oz firm button mushrooms
salt and freshly ground black pepper
few drops Worcestershire sauce
pinch thyme and oregano

Cut off excess fat from cutlets. Skin and thickly slice kidneys. Peel the onions and cut them into thin slices. Slice mushrooms.

Place excess fat from cutlets in a heavy pan. Heat until the fat runs out. Add the cutlets and cook them over a high heat until they are nicely browned on both sides.

Transfer the cutlets to a casserole. Add the kidneys, onions and mushrooms, season with salt and pepper, add the herbs and pour over the stock mixed with the tomato purée and a little Worcestershire sauce.

Cover tightly and cook in a moderate oven (350°F, 180°C, Reg. 4) for 1 hour.

Step 2

To freeze: Leave to cool and skim off any fat from the surface. Transfer to a rigid container, seal, label and freeze.

Step 3

To use

1 thawed Lancashire hot pot recipe
1 lb potatoes
$\frac{1}{2}$ oz melted butter
2 tablespoons finely chopped parsley

Transfer the hot pot to a casserole dish. Peel the potatoes and boil them until almost tender. Drain the potatoes, cut them into slices and arrange the slices on top of the hot pot. Brush with butter and re-heat the hot pot in a moderate oven (350°F, 180°C, Reg. 4) for about $\frac{3}{4}$ hour or until hot through and golden brown on the top.

Sprinkle with parsley before serving.

Lamb patties

(*Serves 6*)

1 large onion
2 oz butter
2 lb lean lamb, minced
1 tablespoon tomato purée
$1\frac{1}{2}$ teaspoons dried mixed herbs
freshly ground black pepper
4 oz fresh white breadcrumbs
1 small egg, beaten

Peel and very finely chop the onion. Melt the butter in a frying pan. Add the onion and cook over a low heat until the onion is soft and transparent. Combine the minced lamb, onion and juices from the pan, the tomato purée, mixed herbs and breadcrumbs in a bowl. Season with freshly ground black pepper and mix in the egg. Chill the mixture and then shape into 16 flat patties using lightly floured hands.

To freeze: Free freeze the patties until solid and then pack in polythene bags or a rigid container with a strip of foil or grease-proof paper between each patty. Seal and label.

To use: Fry the patties from the frozen state in some fat or dripping over a medium low heat for about 10 minutes on each side until crisp, well browned and cooked through. Sprinkle with a little salt after cooking.

Variations: Add some finely chopped parsley in place of the dried herbs.

Add some finely chopped red or green peppers or some finely chopped celery.

Flavour the patties with some garlic if you do not intend to freeze them for longer than 1 month.

Add some minced ham to alter the flavour.

Pork casseroles*

(*Serves 4*)

Step 1 Basic pork recipe

1½ lb lean boned pork
2 onions
½ lb carrots
salt, freshly ground black pepper and a pinch of sage
½ pint stock

Cut the pork into thin slices and then into strips about ¾ inch wide. Peel and slice the onions and divide into rings. Peel the carrots and cut them into matchstick strips.

Layer the pork, onions and carrots in a flameproof casserole, pour over the stock seasoned with the salt, pepper and a little sage, cover tightly, bring to the boil and cook in a moderate oven (325°F, 160°C, Reg. 3) for about 1½ hours or until the meat is tender.

Step 2

To freeze: Cool the stew, remove any fat from the surface and pack in a rigid container. Seal, label and freeze.

Step 3

To use: Thaw overnight (or you can thaw in a moderately hot oven (375°F, 190°C, Reg. 5) for about 1 hour), add 3rd stage ingredients and re-heat in a moderately hot oven (375°F, 190°C, Reg. 5) for the time suggested in the recipes.

Hungarian pork with prunes (*Serves 4*)

Step 3 ingredients

1 portion thawed basic pork recipe
1 tin prunes (8 oz)
¼ pint red wine
3 tablespoons sour cream
2 teaspoons paprika pepper

Drain the prunes. Place the prunes in a saucepan with the red wine and paprika pepper and cook over a high heat until the wine is reduced by one-third. Add the prunes and liquid to the thawed casserole and heat through in a moderately hot oven (375°F, 190°C, Reg. 5) for about 1 hour until hot through. Sprinkle with finely chopped parsley before serving.

Chinese pork (*Serves 4*)

Step 3 ingredients

1 thawed pork recipe
1 large green pepper
2 oz firm button mushrooms
2 oz bean shoots
1 tablespoon soy sauce
1 tablespoon soft brown sugar
1 tablespoon cider vinegar
1 tablespoon cornflour

Remove the core and seeds of the pepper and cut into thin strips. Add the pepper to the pork casserole and heat through in a moderately hot oven (375°F, 190°C, Reg. 5) for about 1 hour until hot through. Add the mushrooms, cut into thin slices, and the beanshoots and continue to cook in the oven for a further 15 minutes. Drain off the liquid from the casserole and place the casserole ingredients on a heated serving dish. Keep the ingredients warm while making the sauce.

Mix the cornflour to a smooth paste with the soy sauce, brown sugar and vinegar. Add the liquid from the casserole and heat over a medium heat, stirring all the time, until the sauce has thickened and is smooth and shining.

Pour the sauce over the casserole ingredients.

Pork with walnuts and apricots (*Serves 4*)

Step 3 ingredients

1 thawed basic pork recipe
1 tin (15 oz) apricots
2 oz shelled walnuts, halved
¼ pint dry cider
4 spring onions
½ tablespoon vegetable oil

Drain the apricots. Put the cider into a small saucepan, bring to the boil and cook over a high heat until the cider is reduced by half.

Add the apricots, walnuts and cider to the basic pork recipe and heat through in a moderately hot oven (375°F, 190°C, Reg. 5) for 1 hour or until hot through.

Cut the spring onions into half lengthwise and cook them in a little vegetable oil for about two minutes until just softened.

Serve the pork with the spring onions arranged on top.

Burlington casserole (*Serves 4*)

Step 3 ingredients

1 thawed pork recipe
juice and grated rind of ½ lemon
½ oz butter
½ oz flour
1½ teaspoons made English mustard

1 small tin tomatoes (14 oz)
1 tin red kidney beans (15 oz), drained
pinch mixed herbs
few drops Worcestershire sauce

Reheat the casserole in a moderately hot oven (375°F, 190°C, Reg. 5) for 1 hour until hot through. Drain off the liquid from the casserole.

Melt the butter in a large saucepan. Add the flour and mix well. Gradually blend in the liquid from the casserole stirring continually until the sauce is thick and smooth. Add the mustard, lemon juice and rind and the tomatoes. Mix in the herbs and Worcestershire sauce, bring to the boil and simmer for 10 minutes. Add the casserole ingredients and beans and heat through. Check seasoning and sprinkle over some finely chopped parsley before serving.

Pork chops with devilled sauce (*Serves 4*)

4 pork chops
2 oz plain flour
1 pint stock
2 tablespoons soft brown sugar
2 tablespoons vinegar
1 tablespoon lemon juice
1 tablespoon dry mustard
1 tablespoon Worcestershire sauce
salt and freshly ground black pepper
1 oz raisins

Trim off excess fat from chops. Put the excess fat in a heavy frying pan and heat over a moderate heat until the fat runs out. Remove excess fat, add the chops to the juices in the pan and cook over a high heat until the chops are brown on both sides. Remove the chops to a shallow baking dish. Add the flour to the juices in the pan and cook over a high heat, stirring, until the flour has browned. Gradually blend in the stock, stirring continually over a medium high heat until the sauce is thick and smooth. Add the sugar, mustard, vinegar, lemon juice and Worcestershire sauce, mix in the raisins and season with salt and pepper.

Pour the sauce over the chops and cover the dish with foil. Bake in a moderate oven (350°F, 180°C, Reg. 4) for 1 hour.

To freeze: Cool the dish, cover it with fresh foil, pack in a polythene bag, seal, label and freeze.

To use: Thaw the dish overnight (or thaw it in a moderately hot oven (375°F, 190°C, Reg. 5) for about 1 hour until thawed through) and re-heat in a moderate oven (350°F, 180°C, Reg. 4) for about 30 minutes or until hot through.

Beef casseroles*

Freeze this starter beef casserole when you can buy stewing beef cheaply. Make eight quantities of the casserole at one time.

Step 1 *(8 servings)*

flour
3 lb lean beef (1·35 kg)
salt and freshly ground black pepper
pinch allspice
3 large onions
1 lb carrots
3 oz dripping
2 pints rich beef stock
bouquet garni

Trim off all fat and gristle from the meat and cut it into 1 inch cubes. Season some flour with salt and freshly ground black pepper and a pinch of allspice. Roll the meat in the seasoned flour until it is well coated on all sides.

Peel and chop or slice the onions. Peel the carrots and cut them into small dice.

Heat the dripping in a heavy saucepan. Add the onions and cook over a low heat until the onions are soft and transparent. Remove the onions from the pan with a perforated spoon, raise the heat, add the meat and cook, stirring every now and then, until the meat is well browned on all sides. Gradually mix in the stock, stirring continually over a medium high heat until the sauce for the casserole is thick and smooth. Add the onions, carrots and bouquet garni, bring to the boil and simmer slowly for 1½ hours or transfer to a casserole, cover tightly and cook in a moderate oven (350°F, 180°C, Reg. 4) for 1½ hours.

Step 2
To freeze: Remove the bouquet garni, cool, pack in rigid plastic containers, seal, label and freeze. Do not freeze for longer than 3 months.

Step 3
To use: Turn the frozen stew into a heavy saucepan and heat slowly until the stew is thawed.

Hasselburg stew (*Serves 4*)

Step 3 ingredients

4 servings starter beef casserole, thawed
¼ pint red wine
1 clove garlic
1 oz shelled walnuts
1 oz butter
grated peel and juice of 2 oranges
1 tablespoon flour

Chop the walnuts. Crush the peeled garlic through a garlic press. Add the wine, garlic, walnuts, orange juice and half the orange peel to the thawed beef stew, bring to the boil and simmer for 30 minutes or until the beef is really tender. Beat the butter with the flour until a smooth paste is formed. Add the butter and flour mixture to the stew and stir until the sauce is thickened and shining. Check seasoning, turn the stew into a serving dish and sprinkle over the remaining orange peel before serving.

Beef casserole with beans (*Serves 4*)

Step 3 ingredients

1 recipe starter beef casserole, thawed
½ teaspoon Worcestershire sauce
1 tablespoon tomato purée
2 cloves garlic
8 oz frozen French beans
1 tablespoon finely chopped parsley

Crush the garlic through a garlic press. Add the Worcestershire sauce, tomato purée, garlic and French beans to the beef stew, bring to the boil, cover tightly and simmer the stew for 30 minutes or until the beef is really tender. Mix in the parsley, check the seasoning and serve.

Beef goulash (*Serves 4*)

Step 3 ingredients

1 recipe starter beef casserole, thawed
1 large tin tomatoes (15 oz)
1 small tin pimentos (7 oz)
4 potatoes
2 teaspoons paprika pepper
¼ pint red wine
1 tablespoon tomato purée

Roughly chop the tomatoes. Drain the pimentos and cut them into thin slices. Peel the potatoes and cut them into quarters. Add the tomatoes, pimentos, wine, tomato purée, potatoes and paprika to the basic stew. Bring to the boil, cover tightly and simmer for 30 minutes, stirring every now and then, until the meat is tender. Check seasoning and serve very hot with boiled rice.

Beef Burgundian (*Serves 4*)

Step 3 ingredients

1 starter beef casserole, thawed
6 oz streaky bacon
2 cloves garlic
6 oz firm button mushrooms
1 oz butter
1 tablespoon flour
¼ pint red wine
1 tablespoon chopped parsley

Remove the rind and cut the bacon rashers into 1 inch pieces. Peel and finely chop the garlic. Cook the bacon, without extra fat, over a medium high heat until the fat has oozed out. Lower the heat, add the garlic and mushrooms and cook over a low heat for three minutes. Remove the bacon, garlic and mushrooms from the pan with a perforated spoon and add them to the stew. Add the butter to the fat in the pan and mix in the flour. Add the fat and flour mixture to the stew and stir over a high heat until the sauce comes to the boil and is smooth. Stir in the red wine, cover and simmer for one hour until the meat is really tender. Stir in the parsley just before serving.

Chilli stew (*Serves 4*)

This is for those who like their dishes 'hot'.

Step 3 ingredients

1 starter beef casserole, thawed
2 green peppers
8 stuffed olives
1 clove garlic
2 tablespoons olive or vegetable oil
¼ teaspoon chilli powder or ¼ teaspoon chilli sauce
1 tin kidney beans (15 oz)
2 tablespoons finely chopped parsley

Remove the core and seeds of the peppers and finely chop the flesh. Peel and finely chop the garlic. Slice the olives. Heat the oil in a small saucepan. Add the peppers and garlic and cook over a low heat until the oil is absorbed. Add the peppers and garlic to the stew with the sliced olives and chilli powder or sauce. Bring to the boil, cover tightly and simmer for 30 minutes. Drain the beans, add them to the stew, check seasonings and continue to cook for a further 15 minutes or until the meat is really tender. Serve with rice and sprinkled with finely chopped parsley.

Carbonnades of beef* (*Serves 6*)

Step 1

2½ lb lean stewing beef
flour
salt and freshly ground black pepper
3 oz dripping or lard
1 tablespoon vegetable oil
4 onions
½ lb carrots
1 tablespoon soft brown sugar
½ pint dark ale
¾ pint strong stock
1 tablespoon tomato purée
1 teaspoon French Dijon mustard
2 teaspoons white wine vinegar
pinch allspice
bouquet garni
3 bay leaves

Cut the steak into 1 inch cubes. Peel and chop the onions. Peel and roughly chop the carrots. Season some flour with salt and pepper and coat the meat in the seasoned flour on all sides. Melt the dripping or lard with the oil in a heavy frying pan. Add the meat and cook over a high heat until the meat is well browned on all sides. Remove the meat with a slotted spoon to a casserole dish. Add the onions and carrots to the juices in the pan and cook over a low heat, stirring, until the onions are soft and transparent. Add the ale and stock, bring to the boil. Mix in the sugar, tomato purée, mustard and vinegar. Add a pinch of allspice. Bring to the boil stirring well and pour the liquid and vegetables over the meat. Add the bouquet garni and bay leaves, cover tightly and cook in a moderate low oven for 2½ hours (325°F, 160°C, Reg. 3) or until the meat is tender.

Step 2

To freeze: Cool the casserole, transfer to a rigid container, seal, label and freeze.

Step 3

To use

4 oz butter
1 small French loaf
2 cloves garlic
2 tablespoons finely chopped parsley

Thaw the casserole overnight and transfer to a casserole dish. Heat through in a moderate oven (350°F, 180°C, Reg. 4) for about 40 minutes until hot through.

Crush the garlic through a garlic press and mix it with the butter. Cut the bread into slices and spread the slices with the garlic butter. Place the slices, buttered side up, on top of the casserole and return to a moderate oven (350°F, 180°C, Reg. 4) for 15 minutes or until the slices are crisp.

Sprinkle with parsley before serving.

Jugged beef* (*Serves 6*)

A dish that uses inexpensive shin beef but which has the flavour of a much more expensive dish.

Step 1

2 lb shin beef
2 tablespoons plain flour
salt and freshly ground black pepper
pinch cinnamon and allspice
2 oz dripping
2 large onions
4 cloves
pinch sugar
½ pint beef stock
½ pint red wine
¾ lb carrots
grated rind of ½ lemon
2 celery sticks
2 bay leaves

Cut the beef into very thin slices and then into strips about 1 inch wide. Peel and slice one of the onions. Peel the second onion and stick the cloves into it. Peel the carrots and cut them into matchstick strips. Slice the celery.

Coat the meat with the seasoned flour.

Heat the dripping in a large heavy pan. Add the slices of beef, a few at a time, and cook over a high heat until the beef is well browned. Remove the beef to a casserole with a slotted spoon. Add the sliced onion to the juices in the pan and cook over a low heat until the onion is soft and transparent. Add the stock, sugar and red wine, a pinch each of cinnamon and allspice, and bring to the boil, stirring all the time.

Add the carrots, celery and lemon rind to the meat in the casserole. Put in the onion with the cloves, pour over the stock and top with the bay leaves. Cover tightly and cook in a moderate oven (325°F, 160°C, Reg. 3) for 2 hours or until the beef is tender.

Step 2

To freeze: Remove the bay leaves and onion stuck with cloves and leave to cool. Skim off any scum from the surface and transfer to a rigid container. Seal, label and freeze.

Step 3
To use

1 recipe thawed Jugged beef
1 lb potatoes
4 oz mushrooms
½ oz butter

Turn the thawed beef into a saucepan and re-heat over a low heat, stirring every now and then until hot through.

Peel the potatoes and cut them into walnut sized pieces. Boil the potatoes until tender and drain well.

Thickly slice the mushrooms and cook them in the butter for about three minutes until just tender.

Add the potatoes and mushrooms to the casserole, test for seasoning and simmer for five minutes, stirring every now and then. Serve with a sprinkling of finely chopped parsley over the surface.

Braised oxtail* (*Serves 6*)

Step 1

2 oxtails
flour
salt and freshly ground black pepper
1 pint well flavoured meat stock
½ pint red wine
bouquet garni
2 large onions
1 lb carrots
4 sticks celery
1½ oz dripping or lard
1 tablespoon lemon juice
6 oz mushrooms
1 tablespoon marmalade

Ask your butcher to chop the oxtails into joints. Season some flour with salt and freshly ground black pepper and roll the oxtail joints in the seasoned flour until well coated. Peel and slice the onions. Peel and chop the carrots. Slice the celery sticks. Thickly slice the mushrooms.

Heat the dripping or lard in a heavy frying pan. Add the oxtails and cook over a high heat until they are browned on all sides. Remove the oxtails to a casserole dish with a slotted spoon and add the onion to the juices in the pan. Cook over a low heat until the onions are soft and transparent. Pour over the stock and wine, bring to the boil, stirring well and pour the liquid and onions over the oxtails. Add the bouquet garni, marmalade and lemon juice, cover tightly and cook in a moderate low oven (300°F, 150°C, Reg. 2) for 2 hours. Add the carrots and celery and continue to cook in a low oven (275°F, 140°C, Reg. 1) for a further 1¾ hours. Mix in the mushrooms and cook for a further 15 minutes.

Step 2

To freeze: Cool and transfer to a rigid container. Seal, label and freeze.

Step 3

To use

finely chopped parsley

rind of 1 orange; blanched, or 6 oz green grapes

Thaw the oxtails overnight, transfer to a casserole and heat through in a moderate oven (350°F, 180°C, Reg. 4) for about 1 hour or until the dish is hot through and the oxtail is tender.

Garnish the dish with finely chopped parsley and blanched orange peel or peeled and seeded grapes.

Jugged hare*

(Serves 4–6)

Step 1

1 hare
hare's blood if available
2 oz plain flour
salt and freshly ground black pepper
4 oz streaky bacon rashers
8 small shallots or onions
bouquet garni
¾ pint stock
grated rind of 1 lemon
4 tablespoons port.

Cut the hare into joints and reserve the hare's liver and blood. Soak the joints and liver in cold water overnight. Remove the rinds from the bacon and chop the rashers. Peel the shallots or onions and leave them whole.

Drain the hare joints, dry thoroughly and toss in seasoned flour with salt and pepper until the joints are well coated.

Fry the bacon without extra fat until the fat runs out. Remove the bacon and add the hare joints to the juices in the pan. Cook the joints over a high heat until they are browned on all sides. Put the hare joints into a casserole with the bacon, onions, lemon rind and the liver cut into small pieces.

Add the stock to the juices in the pan stirring all the time, bring to the boil and stir until the sauce has thickened and is smooth. Mix in the port and hare's blood and pour the gravy over the ingredients in the casserole. Add the bouquet garni, cover tightly and bake in a moderate oven (350°F, 180°C, Reg. 4) for 2 hours or until the hare is tender.

Step 2

To freeze: Cool, pack in a rigid container, seal, label and freeze.

Step 3

To use: Thaw the hare overnight. Transfer to a flameproof casserole and either re-heat over a low flame on top of the stove or in a moderate oven (350°F, 180°C, Reg. 4) for about ¾ hour or until hot through.

Serve the hare with savoury stuffing balls (see page 171).

Minced beef*

Minced beef can sometimes be bought for a very reasonable price but do buy it from a reputable source. Sometimes minced beef is made from all the leftover pieces of beef sold by a butcher and it may be a mixture of good and poor quality meat which may well turn out to be tough. Ideally you should buy your own beef and mince it at home.

With some starter minced beef recipes in your freezer you have access to some really delicious and quickly prepared dishes.

Step 1 *(2–4 servings)*

4 onions
2 oz dripping or 2 tablespoons olive or vegetable oil
2 lb minced beef
2 tablespoons flour
1 tablespoon tomato purée
½ pint rich stock or stock and leftover gravy (or water and stock cube)
salt and freshly ground black pepper
bouquet garni
few drops Worcestershire sauce

Peel and finely chop the onion. Heat the dripping, add the onion and cook over a low heat until the onion is soft and transparent. Raise the heat, add the beef and cook over a high heat until the beef is browned. Add the flour and tomato purée and mix well. Gradually add the stock, stirring continually, until the mixture has thickened and is smooth. Add the bouquet garni and stir in the Worcestershire sauce. Season with salt and pepper, cover and simmer for 30/45 minutes until the meat is almost tender.

Step 2
To freeze: Cool the starter recipe, remove the bouquet garni, pack in two containers, seal, label and freeze.

Step 3
To use: Turn the frozen minced meat into a saucepan and heat over a low heat until thawed through. Use the thawed minced meat recipe for any of the following recipes.

Minced meat cobbler (*Serves 4*)
A delicious savoury recipe with a scone topping flavoured with cheese.

Step 3 ingredients

4 servings minced beef starter
1 small tin tomatoes
1 teaspoon dried mixed herbs
8 oz self raising flour
½ teaspoon salt
2 oz butter
¼ pint milk
3 oz Cheddar cheese, grated
1 small egg

Add the tomatoes to the starter recipe with the dried herbs, and mix well. Put the mince into a shallow baking dish. Sift flour and salt into a bowl, add the butter cut into small pieces and rub the butter into the flour until the mixture resembles fine breadcrumbs. Stir in the milk and mix to form a soft dough. Turn onto a floured board and knead lightly until the mixture is smooth and workable. Roll out to an oblong about 15 in by 7 in (37 cm × 17 cm). Sprinkle over the cheese, roll up like a swiss roll and cut into 12 slices. Place the slices in an overlapping layer on top of the mince. Brush the top with beaten egg and bake in a hot oven (400°F, 200°C, Reg. 6) for 30 minutes or until the scone topping is crisp and golden brown.

Sauce Bolognese (*Serves 4*)
The classic meat sauce that you can serve on top of spaghetti or in a ring of boiled rice. As a snack meal you can serve the sauce

with triangles of crisply fried bread popped onto the top of the minced meat recipe.

Step 3 ingredients

four servings thawed minced beef starter
2 cloves garlic
1 tablespoon tomato purée
1 small tin tomatoes
¼ pint red wine
pinch oregano
pinch sage
few drops Tabasco sauce
2 tablespoons finely grated Parmesan cheese.

Drain the tomatoes reserving the juice and finely chop the tomatoes. Peel and crush the garlic through a garlic press. Add the garlic, tomato purée, chopped tomatoes and tomato juice, red wine, oregano and sage to the minced meat. Mix in the tabasco, bring to the boil and simmer gently until the meat is tender. (If the mixture is too liquid, remove the cover and cook over a high heat until some of the liquid has evaporated.) Check seasoning and serve on top of the spaghetti or as a snack meal with a sprinkle of Parmesan cheese over the top.

Savoury minced meat on toast (*Serves 4*)

A quick freezer recipe for times when you don't want to cook an elaborate meal.

Step 3 ingredients

4 servings thawed minced beef starter
4 oz firm button mushrooms
1 tablespoon finely chopped parsley
2 tablespoons double cream
few drops Worcestershire sauce
toast
butter

Very finely chop the mushrooms. Melt about one ounce of butter in a small frying pan, add the mushrooms and cook over a high heat for 2 minutes. Add the mushrooms to the minced meat, bring to the boil and simmer for 10 minutes or until the meat is absolutely tender. Lower the heat, stir in the cream and flavour with a few drops of Worcestershire sauce. Check seasoning and serve piled on buttered toast with the crusts cut off and sprinkle with parsley.

Chilli con carné (*Serves 6*)

Chilli con carné should be HOT and no messing about. Serve it for a supper party with plenty of ice cold lager and accompany it with rice and a salad of mixed kinds of lettuces, spring onions, cucumber and green peppers so that you get a good contrast to the rich dark red of the chilli con carné.

Step 3 ingredients

1 basic minced meat recipe thawed
1 tablespoon vegetable oil
2 red peppers
2 tablespoons tomato purée
1 small tin tomatoes
½ teaspoon sugar
salt and freshly ground black pepper
1 teaspoon chilli powder
few drops Tabasco sauce
one 15 oz tin red kidney beans

Remove the core and seeds of the red peppers and cut the flesh into thin strips. Roughly chop the tomatoes. Heat the oil in a heavy frying pan. Add the peppers and cook over a low heat, stirring every now and then to prevent sticking until the peppers are soft. Add the thawed mince mixture and mix well. Stir in the tomato purée, tomatoes and sugar, season with salt and freshly ground black pepper and mix in the chilli powder and Tabasco. Bring to the boil, cover tightly and simmer for 30 minutes. Add the kidney beans and their liquid, test for seasoning and heat through.

Hamburgers

The true hamburger is made with meat and meat only, good quality meat using the best minced beef with a proportion of fat in the meat to keep it moist. The better the quality of the meat the better will be the burger. If you really want to make the perfect article then the meat should really be scraped rather than minced as minced meat loses some of its juices in the mincing process – the new food processors on the market which chop rather than mince are ideal for this purpose.

Never add salt to the meat when you are cooking it. Salt draws out the blood from the meat and should always be added after cooking.

The best hamburger meat of all is rump steak minced or chopped with some of the fat from the outside of the steak. Buying ready minced meat can lead to disappointments unless you really know your butcher and trust him not to sell you inferior minced meat.

The mince for hamburgers should be coarse rather than too finely minced.

If you plan to buy meat in bulk and then freeze your hamburgers raw to cook later, make sure that the meat you buy has not been frozen before. Hamburgers can be lightly cooked and then frozen but they tend to lose a lot of flavour this way and I would not really recommend this system.

To make hamburgers

Mince the meat through a coarse mincer, finely scrape it with a sharp knife or fairly finely chop the meat in a food processor. Divide the meat into equal portions. Using floured hands, shape the meat into balls and then flatten the balls to make flat cakes about $\frac{1}{3}$ inch in thickness.

To freeze: Pack the hamburgers in a rigid container with a sheet of foil, waxed paper or Clearfilm between each one to prevent them sticking together. Seal, label and freeze.

To use: Cook the hamburgers straight from the freezer or allow them to thaw out overnight in a refrigerator without unwrapping. Home made hamburgers can be frozen for up to 6 months.

Cooking hamburgers

Whether you use a grill or fry the hamburgers in a pan, the hamburgers must be cooked as quickly as possible and, providing they have a reasonable proportion of fat it should not be necessary to use more than a very small amount of fat when cooking them.

Cook the hamburgers under a high grill (brush them first with a little good vegetable oil) and turn them once during the cooking time.

Or fry the hamburgers in a very hot pan (heated before they are added) in a very thin film of hot oil or fat. Keep the heat as high as possible and turn the hamburgers once during the cooking time.

Cooking time depends on whether you like them rare or well cooked. The size of the hamburger depends on your appetite. You can use 4–8 oz meat for each serving.

Variations: Hamburgers can be given the addition of a range of flavourings as they are being made but remember that many flavourings increase in strength during their freezer life so don't overdo it. Also flavourings can shorten the freezer life of basic ingredients so do not store these hamburgers in the freezer for more than about 2 months.

For a hot hamburger, add some Worcestershire sauce, Tabasco sauce and some very finely chopped parsley. Or try 1 clove finely crushed garlic or some finely chopped spring onions.

Other flavourings: Ground cumin or commercial steak spice or seasoning; a little dried mixed herbs.

Potted meats

Although potted meats can be kept for some time in a refrigerator they have a much longer life in the deep freeze and they make an excellent and quickly prepared alternative to pâtés or terrines as well as making a delicious spread for toast or sandwiches.

Potted ham

8 oz ham
5 oz softened butter
1 teaspoon mixed chopped fresh herbs
pinch nutmeg
ginger
mace and black pepper
few drops lemon juice
2 teaspoons brandy.

Mince the ham and pound it to a smooth paste with the butter. Add the herbs, lemon juice and brandy and season with pepper, ginger, mace and nutmeg.

To freeze: Pack the potted meat into small containers, seal, label and freeze.

To use: Thaw the potted meat overnight in the refrigerator.

Note: The potted meats can be stored in small tubs (i.e. cream tubs) and then turned out to serve.

Variations: Use half ham and half tongue instead of all ham.

Use half ham and half chicken in the place of all ham.

Use beef shin that has been braised until really tender in a rich stock.

Use chicken, game, or some leftover turkey, in the place of the ham.

Poultry

The process of deep freezing has completely revolutionised the role that is played by the chicken and turkey in our every day diet. No longer are chickens a dish only to be enjoyed on festive occasions, now everyone can enjoy poultry at a reasonable price, throughout the year. But inexpensive frozen poultry does have a slight lack of flavour. This isn't really a great problem providing you allow for it in your seasonings, flavourings and sauces.

Chicken liver pâté

(*Serves 6–8*)

8 chicken livers
4 oz butter
1 tablespoon vegetable oil
1 large onion
1 clove garlic
salt and freshly ground black pepper
pinch mace
1 tablespoon very finely chopped or minced parsley
2 teaspoons lemon juice
pinch mixed herbs
1 tablespoon brandy
½ gill double cream

Remove any fibres from the chicken livers. Peel and very finely chop the onion. Crush the garlic. Melt half the butter in a frying pan. Add the livers and cook over a moderately high heat for 4 minutes, turning the livers every now and then. Remove the livers. Add the oil and the onion and garlic to the juices in the pan and cook over a low heat until the onion is soft and transparent. Season the onion with the salt, pepper and mace and add the parsley, lemon juice and dried herbs. Mix well and simmer for 1 minute to bring out the flavour of the herbs.

Combine the remaining butter, livers, brandy and the onion mixture in an electric liquidiser or food processor (or purée them through a fine food mill) and process until the mixture is smooth. Whip the cream until thick and fold it into the chicken liver pâté.

To freeze: Pack in two pots or into small rigid containers. Seal, label and freeze.

To use: Defrost at room temperature for 3 hours and turn out of rigid containers or leave in small pots. Garnish with slices of lemon and serve with plenty of fresh toast and butter.

Chicken casseroles*

(*Serves 4–6*)

Step 1

6 chicken joints
2 tablespoons plain flour
salt and freshly ground black pepper
1 tablespoon cooking oil
1 oz butter
1 large onion
¾ lb carrots
1 bouquet garni
¾ pint chicken stock

Season the flour with salt and pepper and mix well. Coat the chicken joints in the seasoned flour. Peel and chop the onion. Peel the carrots and cut them into small dice.

Heat the oil with the butter, add the chicken joints and cook over a high heat until the chicken joints are golden brown on all sides. Remove the joints with a slotted spoon to a casserole dish. Add the onion to the juices in the pan and cook over a low heat, stirring, until the onions are soft and transparent. Add the stock, bring to the boil and stir well. Pour the onions and liquid over the chicken, add the carrots and bouquet garni, cover tightly, and cook in a moderate oven (350°F, 180°C, Reg. 4) for about 1 hour or until the chicken is tender.

Step 2

To freeze: Cool the casserole, remove the bouquet garni and skim off any fat from the surface. Pack in a rigid container, seal, label and freeze.

Step 3

To use: Thaw the casserole overnight or place in a casserole dish, cover with foil and thaw in a moderately hot oven (375°F, 190°C, Reg. 5) for about 1 hour.

Re-heat thawed casserole in a moderately hot oven for about 30 minutes or until hot through.

Note: You can take the meat off the bones and cube it instead of leaving the joints whole.

Chicken with lemon and tarragon (*Serves 4–6*)

Step 3 ingredients

1 recipe thawed chicken casserole
grated rind and juice of ½ lemon
¼ teaspoon dried tarragon
½ gill double cream

Soak the tarragon in the lemon juice for five minutes. Add the lemon juice, tarragon and lemon rind to the thawed chicken recipe and re-heat in a moderately hot oven (375°F, 190°C, Reg. 5) for 1 hour or until hot through. Remove from the heat, stir in the cream, check the seasoning and serve at once.

Malaysian chicken (*Serves 4–6*)

Step 3 ingredients

1 recipe thawed chicken casserole
1 green pepper
½ oz butter
tin pineapple chunks (8 oz)
1 tablespoon cornflour
2 oz salted peanuts

Drain the pineapple. Remove the core and seeds of the pepper and cut the flesh into thin strips. Melt the butter in a frying pan, add the pepper and cook over a low heat, stirring for 3 minutes. Add the pepper and pineapple to the chicken and heat the dish through in a moderately hot oven (375°F, 190°C, Reg. 5) for about 1 hour.

Mix the pineapple juice and cornflour to a smooth paste. Add the paste to the casserole and mix over a medium low heat, stirring all the time until the sauce has thickened and is smooth. Check seasoning.

Crush the nuts with a rolling pin (or you can chop them in a liquidiser or food processor but the nuts should not be too finely chopped). Roast the nuts in a hot oven for about 1 minute until crisp and golden. Sprinkle the nuts over the casserole and serve at once.

Chicken with angels on horseback in a rice ring (*Serves 6*)

Step 3 ingredients

1 recipe thawed chicken casserole
ingredients for chicken with lemon and tarragon (see page 141)
16 boiled and stoned prunes
8 thin rashers streaky bacon
8 oz long grain rice
2 oz butter
salt and freshly ground black pepper

Cook the rice in boiling, salted water for about 20 minutes until just tender. Drain well. Return to a clean pan, add the butter and season with salt and freshly ground black pepper. Mix well and cook over a moderate heat, stirring until the butter has been absorbed. Press the rice into an oiled ring mould and keep warm.

Follow the instructions for chicken with lemon and tarragon on page 141. Remove the rinds from the bacon rashers and stretch each rasher with the back of a knife. Cut the rashers in half and wrap the halves around each prune. Roast the wrapped prunes in a moderately hot oven (375°F, 190°C, Reg. 5) for about 20 minutes until the bacon is cooked through.

Turn out the ring mould, fill the centre with the chicken and top with the prunes wrapped in bacon.

Chicken paprika (*Serves 4–6*)

Step 3 ingredients

1 thawed recipe chicken casserole
1 tablespoon paprika pepper
1 carton sour cream

Add the paprika to the chicken dish and re-heat in a moderately hot oven (375°F, 190°C, Reg. 5) for 1 hour until hot through. Stir in the sour cream, check seasoning and serve at once with rice and a green vegetable salad.

Crusty cheese topped chicken casserole (*Serves 4–6*)

Step 3 ingredients

1 thawed recipe chicken casserole
3 tablespoons fresh white breadcrumbs
2 oz grated gruyère cheese
½ teaspoon dry mustard
1 oz melted butter

Re-heat the chicken casserole in a moderately hot oven (375°F, 190°C, Reg. 5) for about 30–45 minutes or until hot through.

Mix the breadcrumbs with the grated cheese and dry mustard. Spread this mixture over the top of the chicken, sprinkle with melted butter and put under a hot grill until the topping is crisp and golden brown.

Chicken pie (*Serves 6*)

This is a useful way of stretching a small amount of leftover chicken.

12 oz shortcrust pastry (see page 202)
1½ oz butter
1 oz flour
scant ½ pint chicken stock
¼ pint milk
6–8 oz cooked chicken
4 oz firm button mushrooms
1 small onion
1 green pepper
salt and freshly ground black pepper

Set aside ⅓ of the pastry for the top of the pie. Roll out the remaining pastry to ⅛ inch thickness and line a 8 in flan case with the pastry.

Peel and finely chop the onion. Remove the core and seeds of the green pepper and finely chop the flesh. Slice the mushrooms.

Melt ½ oz butter in a frying pan. Add the onion and green pepper and cook over a low heat until the onion is soft and transparent and most of the butter has been absorbed. Add the mushrooms and toss over a medium high heat for 2 minutes.

Melt 1 oz butter in a saucepan. Add the flour and mix well. Gradually blend in the stock and milk, stirring continuously until the sauce comes to the boil and is thick and smooth. Add the chicken, onion, pepper and mushrooms and season with salt and freshly ground black pepper. Leave to cool.

Fill the flan case with this mixture and top with the remaining pastry, rolled out thinly. Damp the sides of the case and press the edges firmly together. Flute the edges with the back of a fork. Do not cut air vents in the top of the pastry.

To freeze: Free freeze the pie and then pack in a polythene bag. Seal, label and freeze.

To use: Unwrap the pie and cook in a hot oven (425°F, 220°C, Reg. 7) for 15 minutes. Brush the top with beaten egg, cut two vents in the top and return to the oven for a further 30 minutes or until the pie is golden brown and the contents are hot through.

Cassoulet

(*Serves 10–12 or two portions of 5–6*)

This traditional French peasant dish is a popular one in our family. It is very rich and should only be preceded by a light starter. I find it a good dish to serve for parties in the winter and, as all the ingredients are chopped, it makes a good hot buffet dish.

The finishing touch of the finely chopped parsley, butter and garlic is very important.

8 oz dried haricot beans
2 large onions
6 oz streaky bacon in one piece if possible
1 bayleaf and two or three parsley stalks
pinch thyme and sage
1 small duck
8 oz garlic sausage
salt and freshly ground black pepper
1 tablespoon tomato purée
1 small tin tomatoes

Roast the duck in the usual way until tender and leave to cool.

Soak the beans in water overnight and drain well. Remove the rind, as thinly as possible, from the bacon and cut it into thin slices. Peel and roughly chop the onions. Put the beans into a large heavy saucepan with the bacon, chopped onion, parsley stalks, bay leaf, sage and thyme. Cover with cold water, season with salt and pepper, bring slowly to the boil, lower the heat and simmer slowly for 1½ hours or until the beans are tender. Drain the cooking liquid into a separate pan and remove the bay leaf and parsley stalks.

Remove the meat from the duck and cut it into thick strips. Remove the skin from the garlic sausage and cut it into small cubes.

Add the tomatoes and tomato purée to the cooking liquid.

Bring to the boil and simmer for 10 minutes breaking up the tomatoes with a fork. Add the garlic sausage, duck and cooking liquid to the bean mixture, cover tightly and simmer for a further 40 minutes or until the duck is really tender.

To freeze: Cool, divide into two portions, pack in rigid containers; seal, label and freeze.

To use:

3 tablespoons finely chopped parsley
2 cloves garlic
1 oz softened butter

Thaw overnight in a refrigerator. Transfer to a fireproof oven dish. Re-heat in a slow oven (300°F, 150°C, Reg. 2) for about 1½ hours, stirring every now and then until the dish is hot through.

Mix the parsley, garlic and butter to a paste. Add the paste to the hot cassoulet and mix lightly until the butter has melted and the parsley and garlic is well distributed.

Serve the cassoulet with rice and a cabbage recipe – the best of all is spiced red cabbage (see page 161).

Fish

The most important thing when dealing with any fish for the freezer is to make absolutely sure the fish is as fresh as fresh can be. Confine yourself to only freezing fish that has been caught that day and take special care when dealing with shellfish which can be particularly hazardous.

Smoked fish mousse

(Serves 6)

If you feel like really splashing out you can make this mousse from smoked trout or smoked salmon pieces. For less extravagant occasions, use the recipe for a kipper, bloater or smoked mackerel mousse instead. Cook the kippers before using them.

¾ lb smoked fish on the bone
4 oz cottage cheese
1 carton (5 fl oz) sour cream
juice half a lemon
1 tablespoon finely chopped chives
freshly ground black pepper
small pinch of mace and cayenne

Remove the skin and bones from the fish. Flake the fish and pound it to a paste with the cottage cheese. When the mixture is smooth (this can be done in an electric liquidiser or food processor), mix in the chives and blend in the sour cream and lemon juice. Season with freshly ground black pepper, mace and a little cayenne.

To freeze: Pack the mousse in a square, rigid container. Seal, label and freeze.

To use: Thaw the mousse in a refrigerator overnight, turn out and garnish with thin wedges of lemon and some finely chopped parsley. Serve chilled with hot toast.

Devilled crab mousse

(Serves 6)

8 oz crab meat (mixed brown and white)
salt and freshly ground black pepper and a pinch of cayenne
1 oz cream cheese
6 fl oz jellied consommé
1 teaspoon gelatine
juice ½ lemon
1 teaspoon made mustard
few drops Tabasco sauce
few drops Worcestershire sauce
½ gill double cream
2 egg whites

Put the gelatine into a bowl with the lemon juice and 1 tablespoon of consommé and heat over a very low heat, stirring until the gelatine has dissolved.

Mix the cream cheese with the crab meat until smooth. Add the mustard, season with salt, pepper and a pinch of cayenne and add a few drops of Tabasco and Worcestershire sauce. Add the melted gelatine mixture to the consommé and mix well. Blend the consommé with the crab mixture and chill until the mousse base is on the point of setting.

Beat the cream until thick. Whisk the egg whites until stiff. Fold the cream and then the egg whites lightly into the crab mixture and turn into a soufflé dish or into ramekins.

To freeze: Cover with Clearfilm, pack in polythene bags, seal, label and freeze.

To use: Thaw the mousse in the refrigerator overnight. Unwrap the mousse, turn it out and garnish with thin slices of peeled cucumber and thin slices of lemon. Serve with hot toast.

Fish pie

(*Serves 4–6*)

This is another of those dishes where you can ring the changes or alter the ingredients to suit the season or what you happen to have available. Fish pies freeze well (providing you don't include any hard boiled eggs).

1 lb white fish (use cod, coley, haddock etc.)
¼ pint dry white wine or dry cider
1½ oz butter
1½ oz flour
¼ pint milk
4 tablespoons single cream (or you can use more milk)
4 small tomatoes (peeled, seeded and chopped)
4 oz frozen peas
1 tablespoon finely chopped parsley
1 tablespoon finely chopped chives or spring onion tops
3 oz finely grated Cheddar cheese
salt and freshly ground black pepper
small pinch mace
1 lb potatoes
2 tablespoons milk
1 oz butter

Place the fish in a shallow pan. Pour over the cider or wine, season with salt and pepper, bring to the boil, cover and simmer gently until the fish is tender. Remove the fish and discard any skin or bones. Flake the fish. Cook the peas in boiling salted water until tender and drain well.

Heat 1½ oz butter in a saucepan, add the flour and mix well. Gradually blend in the cider or wine and milk, stirring continually over a medium high heat until the sauce is thick and smooth. Add 2 oz cheese to the sauce and stir until the cheese has melted. Add the cream, fish, peas, tomatoes, parsley, mace and chives to the sauce, season with salt and pepper and mix lightly. Turn the fish mixture into a pie dish or foil container.

Cook the potatoes in boiling, salted water until tender. Mash until smooth and beat in 2 tablespoons milk and 1 oz butter. Season with salt and pepper and spread the potatoes on top of the pie. Sprinkle over the remaining cheese.

To freeze: Leave the dish to cool, pack in a polythene bag, seal, label and freeze.

To use: Thaw the dish overnight and then re-heat in a moderately hot oven (375°F, 190°C, Reg. 5) for 30 minutes.
OR put the frozen pie into a moderately hot oven (375°F, 190°C, Reg. 5) and cook for 1 hour or until the pie is hot through and the topping is golden brown.

Fish cakes

(Makes 12 small fishcakes)

Look out for coley which is often very good value indeed; pollock can also be relatively inexpensive when it is in season and so can grey mullet. Ask your fishmonger to skin and fillet the fish if it is on the bone, but make sure you take the trimmings home with you to make into a fish stock that can form the basis of any number of delicious fish soups or sauces.

1½ lb white fish
fish stock or water with a slice of onion and a bay leaf
1 lb potatoes (peeled weight)
2 oz butter
salt, freshly ground black pepper and a little cayenne
1 teaspoon lemon juice
1 tablespoon finely chopped parsley
1 tablespoon finely chopped chives
2 eggs
breadcrumbs

Place the fish in a deep pan with a teaspoon of lemon juice and cover it with cold fish stock or water with a slice of raw onion, a bay leaf and some salt and freshly ground black pepper. Bring to the boil and simmer gently until the fish is cooked enough to flake with a fork. Drain, cool and flake the fish.

Cook the potatoes until tender, drain well and mash until smooth.

Beat in the butter and season well with salt, pepper and cayenne. Mix in the chopped parsley and chives and then mix in the fish. Leave to cool.

Shape the cooled mixture into flat cakes about $\frac{1}{3}$ inch thick and chill the cakes in a refrigerator until firm.

Beat the eggs with a little salt until smooth. Dip the cakes into the beaten egg and coat in breadcrumbs.

To freeze: Spread out the cakes on a shallow tray covered with foil and free freeze until solid. Pack in rigid containers with a slip of foil or greaseproof paper between each cake. Seal, label and freeze.

To use: Fry the frozen fish cakes in mixed butter and oil for about 6–10 minutes on each side until they are crisp, golden brown and hot through. Serve the cakes with a caper, egg and mustard or home made tomato sauce.

Variations: Add some finely chopped, peeled prawns, or crisply fried and crumbled streaky bacon rashers or grated cheese to the mixture.

After cooking the fish cakes add a slice of tomato and some grated cheese to them and put under a grill until the cheese has melted and is bubbling.

Vegetables

Part of the delight and economy of owning a deep freezer is to be able to freeze those vegetables you have grown yourself or to buy in bulk when vegetables are cheap and plentiful and to prepare and freeze them for your own use. Take care, however, not to become a slave to your freezer or to have to spend the summer months eternally podding or slicing vegetables which your family are never really going to enjoy.

Using frozen vegetables

Nearly all vegetables should be cooked straight from their frozen state making frozen vegetables one of the most quickly-served products your freezer has to offer.

Whenever possible top quality frozen vegetables are best cooked in a little butter without extra liquid; this applies especially to frozen peas and beans of all sorts, to spinach, Brussels sprouts and courgettes.

Melt about $1\frac{1}{2}$ oz butter in a saucepan. Add the frozen vegetables (they must be free frozen and not in a solid lump), season with freshly ground black pepper, cover tightly and cook over a medium heat, shaking the pan every now and then, until the vegetables are just tender.

If you do cook your frozen vegetables in water, use small quantities of fast-boiling salted water. Do not overcook vegetables as you will spoil both their texture and flavour and impair their nutritional value. However carefully you freeze your own vegetables there are occasions when they may have lost some of their freshness before freezing so that their flavour may be a little disappointing or you may just be bored with having plain peas or beans etc., yet again. Here are some ideas for jazzing up frozen vegetables from your freezer to give them an extra interest. (See also the blanching and cooking guide on pages 66–75.)

Peas with mint butter

(*Serves 4*)

1 lb frozen peas
3 oz butter
3 tablespoons very finely chopped mint
salt and freshly ground black pepper

Cook peas in boiling, salted water until just tender and drain well. Melt the butter in a saucepan, add the mint and peas, season with salt and pepper and toss over a high heat until butter has melted and peas are hot through.

Spiced spinach

(*Serves 4*)

1 lb frozen spinach
3 oz butter
1 teaspoon lemon juice
½ teaspoon mixed ground nutmeg and cinnamon
tiny pinch ground cloves
salt and freshly ground black pepper

Bring half a gill of salted water to the boil. Add the spinach and cook over a high heat, breaking up the spinach with a fork, until it is just tender. Drain really well.

Melt the butter in a saucepan. Add the spices, lemon juice and seasonings. Add the spinach and turn it gently in the butter over a low heat until hot through and coated with the spiced butter.

Sweet and sour beans

(*Serves 4*)

1 lb sliced green French or runner beans
3 oz butter
1 medium onion
1 tablespoon brown sugar
2 tablespoons white wine vinegar
salt and freshly ground black pepper

Cook the beans in boiling salted water until just tender but still crisp, and drain well. Peel and very finely chop the onion. Melt the butter in a saucepan, add the onion and cook over a low heat for about five minutes until the onion is soft and transparent. Add the sugar, vinegar and seasonings, and stir until the sugar has melted. Add the beans and toss them in the sweet and sour sauce until the beans are coated with the sauce.

Curried Brussels sprouts

(*Serves 4*)

1 lb Brussels sprouts
3 oz butter
1 teaspoon Madras curry powder
1 oz flaked almonds
salt and freshly ground black pepper

Cook the Brussels sprouts in boiling salted water until just tender and drain them very well.

Bake the flaked almonds in a hot oven until they are golden and crisp. Leave to cool. Melt the butter in a saucepan, add the curry powder and cook over a low heat, stirring for 2 minutes. Add the almonds and season with salt and pepper. Add the Brussels sprouts and toss them in the butter until they are coated with the curry sauce and hot through.

Broccoli with butter and horseradish

(Serves 4)

This dish has a spectacular taste but you must use fresh horse-radish not the commercial horseradish sauce.

1 lb frozen broccoli
4 oz butter
2 tablespoons freshly grated horseradish
salt and freshly ground black pepper

Cook the broccoli in boiling salted water until just tender. Drain well. Melt the butter in a saucepan, add the grated horseradish and season with salt and freshly ground black pepper. Toss the broccoli lightly in the butter until it is well coated and hot through.

This is delicious with roast beef.

French beans Mimosa

(Serves 4)

1 lb frozen whole French beans
3 oz butter
1 teaspoon lemon juice
salt and freshly ground black pepper
2 hard-boiled eggs

Cook the beans in boiling salted water until tender but still crisp and drain well.

Halve the hard boiled eggs and remove the yolks. Finely chop the egg whites. Melt the butter in a saucepan, add the lemon juice and season with salt and freshly ground black pepper. Add the whites of the eggs and the beans and toss over a medium heat until the beans are coated with the sauce and hot through. Pile them onto a serving dish and top with the hard boiled egg yolks rubbed through a coarse sieve.

Mixed vegetables with a crisp topping

(*Serves 4–6*)

1 lb mixed vegetables (or you can use cauliflower, carrots, beans etc.)
4 oz butter
4 oz fresh white breadcrumbs
2 cloves garlic
4 oz Cheddar cheese, grated
1 oz grated Parmesan cheese
salt and freshly ground black pepper

Cook the vegetables in a little boiling, salted water until just tender but still crisp and drain well.

Peel and crush the garlic.

Melt the butter in a saucepan, add the crumbs and fry until the crumbs are crisp and golden brown. Add the garlic and Cheddar cheese and season with salt and freshly ground black pepper.

Lightly mix the vegetables with the crumb mixture and turn into a lightly greased baking dish. Sprinkle over the Parmesan and put under a grill until the cheese is beginning to melt.

This goes very well with cold cuts of meat or with cold roast chicken.

Aubergine ratatouille

(*Serves 4*)

There are many varieties of ratatouille most of which are equally good served hot or cold. This is one to make when aubergines are well in season and not too expensive.

3 medium aubergines
3 tablespoons olive or good vegetable oil
1 large onion
1 lb tomatoes (or use tinned tomatoes)
1 tablespoon tomato purée
2 cloves garlic
salt and freshly ground black pepper
pinch basil and oregano

Peel and dice the aubergines into fairly small cubes. Sprinkle them with salt and leave to 'sweat' in a colander for 30 minutes. Peel fresh tomatoes by covering them with boiling water for one to two minutes and then sliding off the skins. Halve the tomatoes, scoop out the seeds and roughly chop the flesh. Peel and finely chop the onion. Peel and crush the garlic.

Pat the aubergine cubes dry on kitchen paper. Heat the oil in a heavy saucepan. Add the aubergines and cook over a high heat until they are golden. Remove the aubergines from the pan with a perforated spoon. Add the onion and garlic to the remaining oil in the pan and cook over a low heat until the onion is soft and transparent. Add the tomatoes and tomato purée to the onion. Season with salt and pepper and mix in the herbs. Bring to the boil, cover and simmer for 20 minutes – the mixture should not be too liquid; if there is too much tomato juice, raise the heat and stir until it is reduced. Add the aubergines and continue to simmer for a further 15 minutes.

To freeze: Leave to cool, pack in a rigid container, seal, label and freeze.

To use: Extra ingredients: 4 tablespoons fresh white bread-crumbs; olive or vegetable oil. Thaw at room temperature for about 3 hours, turn into a shallow lightly greased baking dish, sprinkle over the crumbs and dribble over a little olive or vegetable oil. Bake in a moderately hot oven (375°F, 190°C, Reg. 5) for 20 minutes or until the dish is hot through and the topping is crisp and golden brown.

Note: You can freeze the mixture in a baking dish and cook from frozen in a moderately hot oven (375°F, 190°C, Reg. 5) for about 45 minutes until hot through and golden brown.

Variations: The crumbs can be mixed with some grated cheese to make a savoury supper dish.

Do not freeze for longer than 2 months.

Courgette ratatouille (*Serves 4*)

If you grow your own courgettes you will inevitably find that you always end up with too many of them and, delicious as these small marrows are when they first appear, you can get very bored of their rather insipid flavour especially when they grow too large. Combining the courgettes with tomatoes, onion and green peppers can make a really exciting dish to serve as a vegetable or as a cold starter.

1 lb courgettes
1 large onion
1 lb tomatoes
3 tablespoons olive oil or a good vegetable oil
2 tablespoons tomato purée
1 large green pepper
salt and freshly ground black pepper
pinch chervil

Cut the courgettes into ¼ inch thick slices. Peel the tomatoes by covering them with boiling water for a minute or two and then sliding off the skins. Halve the tomatoes and remove the seeds with a spoon. Roughly chop the tomatoes; peel and slice the onion; remove the core and seeds of the pepper and cut the flesh into very thin strips.

Heat the oil in a heavy saucepan. Add the onion and pepper and cook over a low heat until the onion is soft and transparent. Add the courgettes and tomatoes, stir in the tomato purée and season with salt and freshly ground black pepper. Bring to the boil, mix in the chervil and cover tightly. Simmer the mixture for 30 minutes removing the cover for the last five minutes of cooking time if there is too much liquid.

To freeze: Cool, spoon into a rigid container, seal, label and freeze.

To use: Turn the frozen mixture into a heavy bottomed saucepan and cook over a low heat until the mixture is thawed and hot through. Stir during thawing time to prevent sticking.

Storage: Do NOT freeze for longer than 6 months.

Leeks in ham blankets (*Serves 4*)

A good savoury dish that freezes well.

8 thin leeks
salt
8 thin slices cooked ham (or use 4 fat leeks and four thicker slices of ham)
1 oz butter
2 tablespoons flour
½ pint milk
6 oz well flavoured Cheddar cheese
1 teaspoon made English mustard
white pepper

Clean the leeks, cover them with cold water, add a pinch of salt, bring to the boil, cover and cook until the leeks are just tender. Drain the leeks (reserving ¼ pint of the water they were cooked in) by placing them upside down in a colander so that all the water runs out of them. Wrap the leeks in the ham slices and arrange them in a shallow buttered baking dish or foil dish.

Melt the butter in a small saucepan. Add the flour and mix well. Gradually add the milk, stirring continually over a medium high heat until the sauce is thick and smooth. Add the water in which the leeks were cooked and continue to stir until the sauce is smooth. Add 4 oz cheese and the mustard, season with salt and white pepper and cook over a low heat, stirring until the cheese has melted. Pour the sauce over the ham parcels and sprinkle over the remaining cheese.

To serve now: Put the dish under a medium high grill and cook until the top is bubbling and golden and the dish is hot through.

To freeze: Pack the dish in a polythene bag, seal, label and freeze.

To thaw and serve: Thaw at room temperature for about 4 hours. Re-heat in a moderate oven (350°F, 180°C, Reg. 4) for 20 minutes. Then finish under a grill until the topping is bubbling and golden and the dish hot through.

Duchesse potatoes

(Serves 4)

These attractive potatoes can be made in advance and frozen to produce a quick vegetable dish for special occasions.

1½ lb potatoes
1 oz butter
salt and white pepper
pinch ground nutmeg
1 egg yolk

Peel the potatoes and boil them until tender in salted water. Drain well and mash the potatoes until smooth. Return the potatoes to a clean pan and stir over a medium heat to get rid of any excess moisture. Add the butter, season with salt and pepper, add a pinch of nutmeg and beat until the butter has melted. Beat in the egg yolk and pipe the potatoes through a large rose nozzle onto a lightly greased baking sheet lined with foil.

To freeze: Free freeze the potatoes until solid. Peel them off the foil and pack them in polythene bags. Seal, label and freeze.

To use: Extra ingredients: 1 small egg, beaten. Place the frozen potatoes on a lightly greased baking sheet and bake in a hot oven (400°F, 200°C, Reg. 6) for 10 minutes. Brush with the beaten egg and return to the oven for a further five minutes until golden brown and crisp.

Variations: Add a little grated Parmesan cheese to the potato mixture and season it with a pinch of cayenne pepper in the place of the nutmeg.

Pipe the potatoes into nests and fill them when cooked with peas or other cooked vegetables.

Storage: Do not freeze the potatoes for longer than 2 months.

Spiced red cabbage

(*Serves* 6)

The red cabbage season is really relatively short so make this spicy vegetable dish when the cabbages are at the height of their season and easy to find.

1 medium red cabbage
1 onion
1 oz lard or dripping
2 cooking apples
1 tablespoon soft brown sugar
2 tablespoons red wine vinegar
½ teaspoon caraway seeds
pinch nutmeg and cinnamon

Remove the tough outer leaves of the cabbage, cut out the stalk and finely shred the rest. Peel and chop the onion. Peel, core and grate the apples.

Heat the lard in a heavy pan. Add the onion and cook over a low heat until the onion is soft and transparent. Add half the cabbage, cover with apples, sprinkle with the sugar and caraway seeds, season with salt, pepper, cinnamon and nutmeg, cover with the remaining cabbage and pour over the red wine vinegar. Cover with a sheet of greased greaseproof paper and then with a tight fitting lid and cook over a low heat for 1½ hours.

Note: The ingredients can be layered and cooked in a slow oven (325°F, 160°C, Reg. 3) for 2½ hours instead of on top of the stove.

To freeze: Cool, pack in a rigid container, seal, label and freeze.

To use: Thaw at room temperature for about 4 hours or overnight in the refrigerator and heat through gently in a heavy bottomed pan until thoroughly hot. Stir to prevent sticking.

Stuffed green peppers

(*Serves 2*)

2 large peppers
1 medium onion
1 tablespoon vegetable oil
3 oz long grain rice
1 clove garlic
8 oz raw minced beef
¾ pint stock
2 teaspoons tomato purée
salt and freshly ground black pepper
2 tomatoes
2 tablespoons fresh breadcrumbs
1½ oz Parmesan cheese

Halve peppers and remove the core and seeds. Parboil peppers in boiling, salted water for 5 minutes and drain well.

Peel and finely chop the onion. Crush the garlic. Peel the tomatoes (cover them with boiling water for 1 minute and then slide off the skins), remove the core and seeds and chop the flesh.

Heat the oil, add the onion and garlic and cook over a low heat until the onion is soft and transparent. Add the meat and cook over a moderately high heat until the meat has browned. Add the rice and mix well. Gradually blend in the stock and tomato purée and mix in the tomatoes. Season with salt and pepper and cook over a medium heat until the rice is cooked and most of the liquid has been absorbed.

Fill the peppers with the stuffing. Sprinkle the mixed breadcrumbs and cheese on top of each pepper.

To freeze: Cool, free freeze the peppers. Wrap each one in Clearfilm, overwrap in polythene bags, and seal, label and freeze.

To use: Unwrap the frozen stuffed peppers and place them in a lightly greased casserole, pour over ¼ pint stock, dribble a little oil over the top of the peppers, and bake in a moderate oven (350°F, 180°C, Reg. 4) for about 1 hour until the peppers are hot through and the topping is a crusty golden brown.

Sauces and Stuffings

When you are in a hurry it can be wonderfully rewarding to know that you have meat or tomato sauces for spaghetti or quick dishes at hand in the freezer. And if you want to prepare ahead for Christmas, or some other festive occasion, it can save time later to make up a stuffing for your poultry and have it ready made, in the freezer, for the time when it will be required. Store sauces in small containers and freeze stuffings separately from the bird they are to be served with.

Freezing sauces

I really do not understand why so many cookery books recommend that you should freeze sauces like white sauces and their variations when it takes only a short time (less than it would take to thaw these sauces) to make them from scratch. It does, however, make sense to freeze those sauces that are made with seasonable ingredients and some that take quite a time to prepare, such as a meat sauce, and which can be made in bulk.

Tomato sauces are excellent freezers and can be made with second quality tomatoes when these are plentiful; an excellent meat sauce can be made from second quality minced meat and other sauces from seasonable ingredients such as cucumbers, gooseberries and plums.

Freezing sauces: Freeze sauces in fairly small quantities (you can freeze them in ice cube containers so that, instead of using the mixture as a sauce, it can be added to soups, stews or casseroles to give extra flavour). Pack the sauces into small rigid containers or small polythene bags, seal, label and freeze.

Thawing sauces: Most sauces can be thawed by being turned into saucepans straight from the frozen state and gently heated over a low flame until thawed and hot through.

Adding cream to sauces: It is advisable to add cream to sauces after they have been thawed and re-heated to prevent the likelihood of curdling.

Note: The storage time for most sauces is 3–6 months. It is preferable to add garlic when re-heating sauces.

Apple sauce

3 large cooking apples	1 oz butter
1 tablespoon granulated sugar	salt and white pepper
1 tablespoon water	

Peel, core and roughly chop the apples. Place the apples in a saucepan with the water and sugar and season with a little salt and pepper. Bring to the boil, cover and simmer very gently until the apples are soft. Add the butter and purée the apples until smooth.

To freeze: Cool, pack in rigid containers, seal, label and freeze.

To use: Turn into a lightly buttered saucepan and heat gently, stirring every now and then, until thawed through (or leave to thaw overnight).

Variations: For a more interesting sauce, add a pinch of mixed herbs, or 1 tablespoon mint jelly, or some grated orange rind and a little marmalade to the apples while cooking them.

Tomato sauce

Tomato sauces go well with gammon, tongue, chicken, fish, lamb and hamburgers. They can also be re-heated with stock to make a quick and excellent soup.

Italian home-made tomato sauce

This sauce must either be put through a sieve or fine food mill or the tomatoes must be peeled before making the sauce which can then be liquidised. The first is the best method as the skins do add flavour to the sauce and the pips will remain in the second method. If you use the second method add the skins to stocks to give additional colour and flavour.

1 small onion
1 stick celery
1 teaspoon fresh basil leaves
1 bunch parsley
4 tablespoons olive or vegetable oil
salt and freshly ground black pepper
2 lb ripe tomatoes
2 bay leaves

Peel and finely chop the onion. Finely chop the celery including the leaves. Mince or very finely chop the basil and parsley. Chop the tomatoes.

Heat the oil in a saucepan. Add the onion and celery and cook over a low heat until the onion is soft and transparent. Add the tomatoes and herbs, season with salt and freshly ground black pepper, bring to the boil and cover tightly. Simmer for 45 minutes and purée through a fine sieve or food mill.

To freeze: Cool, divide into portions, pack in small rigid containers or polythene bags, seal, label and freeze.

To use: Turn the frozen sauce into a saucepan which has been rubbed with a thin film of oil and heat gently, stirring every now and then, until the sauce is thawed and hot through.

Note: A crushed clove of garlic can be added when re-heating the sauce.

Tomato sauce with Ham

This is a more substantial tomato sauce which does well as a topping for a pasta dish. I also serve it cold as a sandwich spread or hot as a filling for small pastry tartlets to serve as a first course. You can use it as a flavouring ingredient for baked eggs or as a Pizza topping.

1½ lb tomatoes
¼ pint chicken stock
1 small shallot or onion
2 teaspoons finely chopped parsley
pinch thyme and oregano
2 bay leaves
salt and freshly ground black pepper
1 oz butter
½ oz flour
pinch sugar
4 oz ham, minced

Roughly chop the tomatoes. Peel and finely chop the shallot or onion. Combine the tomatoes with the shallot or onion, the herbs and sugar and mix in the stock. Bring to the boil, cover and simmer for 1 hour. Rub the sauce through a sieve or a fine food mill.

Melt the butter in a clean saucepan. Add the ham and cook over a low heat for 2 minutes. Add the flour and mix well. Gradually blend in the sauce, stirring continually over a medium high heat until the sauce is thick.

To freeze: Cool, divide into portions, pack in small rigid containers or polythene bags, seal, label and freeze.

To use: Turn the frozen sauce into a heavy saucepan and re-heat over a low heat, stirring every now and then, until the sauce is hot through.

Note: Add a little extra stock if you wish to thin the sauce or to stretch it in emergencies.

Tomato sauce with red wine

A slightly more robust sauce.

2 lb tomatoes
1 small onion
1 small bunch parsley
1 oz lard
$1\frac{1}{2}$ tablespoons olive or vegetable oil
1 oz butter
1 oz flour
$\frac{1}{4}$ pint chicken stock
1 tablespoon lemon juice
1 teaspoon sugar
4 tablespoons red wine
salt and freshly ground black pepper

Roughly chop the tomatoes. Peel the onion and mince it with the parsley leaves and stalks. Heat the oil and lard, add the tomatoes, onion and parsley, season with salt and pepper, bring to the boil cover and simmer gently for 30 minutes adding a little water if the mixture gets dry. Purée the mixture through a sieve or fine food mill.

Melt the butter in a clean saucepan. Add the flour and mix well. Gradually add the stock and red wine stirring continually over a medium high heat until the sauce is thick and smooth. Add the tomato purée, sugar and lemon juice, mix well and continue to simmer for three minutes.

To freeze: Cool, pack in small rigid containers or polythene bags, seal, label and freeze.

To use: Turn the frozen sauce into a saucepan and heat gently, stirring every now and then, until the sauce has thawed and is hot through. Check seasoning before serving.

Note: A crushed clove of garlic can be added to the sauce when re-heating.

Barbecue sauce

Make this in bulk when tomatoes are cheap and use it served with roast, fried, grilled or barbecued foods or use it as a basting sauce for barbecued ingredients.

½ pint home-made Italian tomato sauce (see page 165)
pinch chilli powder
2 teaspoons brown sugar
1 teaspoon mixed herbs
2 tablespoons cider vinegar
salt and pepper

Combine all the ingredients in a saucepan, cover tightly and simmer gently, stirring every now and then for 15 minutes.

To freeze: Cool the sauce, pack in rigid containers, seal, label and freeze.

To use: Turn the frozen sauce into an oiled saucepan and heat through over a low heat, stirring every now and then until the sauce is hot through.

Note: Some crushed garlic can be added when re-heating the sauce but it should not be included before the sauce is frozen as this would limit the freezer life of the sauce.

Cranberry sauce *(Serves 6)*

Cranberries come into the shops in November and this delicious sauce served with turkey at Christmas can be made in advance and frozen. The sauce also goes well with pork, bacon dishes and sausages. You can, of course, also use frozen cranberries for the sauce.

1 lb fresh cranberries
¼ pint water
6 oz sugar

Place the cranberries in a heavy saucepan with the water, bring to the boil and simmer gently for 8 minutes. Remove from the heat, add the sugar and stir until the sugar has dissolved.

To freeze: Leave to cool, pack in a rigid container, seal, label and freeze.

To use: Transfer the frozen sauce to a saucepan and heat over a low heat until the sauce is thawed and hot through. The sauce can be served hot or cold.

Gooseberry sauce

This is traditionally served with mackerel but it has always seemed to me that this is in fact a sauce which could well enhance all sorts of other dishes from plainly grilled or boiled fish to roast chicken or goose.

Make plenty of the sauce when gooseberries are cheap and in season.

$\frac{3}{4}$ lb green gooseberries (they can be very unripe)
$\frac{1}{4}$ pint sweet cider
1 tablespoon cornflour
1–2 tablespoons sugar
1 oz butter
salt and white pepper

Top and tail the gooseberries. Combine 1 tablespoon of cornflour with one tablespoon of cider and mix to a smooth paste.

Put the gooseberries in a saucepan with the remaining cider and cook for about 15 minutes until tender. Purée the gooseberries through a fine food mill, a sieve or in an electric liquidiser. Return the purée to a pan, add the cornflour mixture, butter and sugar and cook over a low heat, stirring occasionally, until the mixture is thickened and smooth. Taste for sweetness and season with a little salt and pepper.

To freeze: Cool, pack in rigid containers, seal, label and freeze.

To use: Turn the frozen sauce into a saucepan and heat through over a low heat until the sauce is hot through. Serve the sauce hot or warm.

Stuffing

Christmas is such a busy time that it helps to prepare as much as you can in advance so that you have the maximum of time to enjoy yourself over the holiday. You can make your stuffings as long as 1 month ahead and have them ready to put into the turkey on Christmas Eve.

Here are two traditional stuffings to cope with both ends of the turkey.

Sausagemeat stuffing

1 lb sausagemeat
2 oz fat streaky bacon
liver and heart from the turkey
1 large onion
2 oz fresh white breadcrumbs
salt and freshly ground black pepper
2 oz butter
pinch sage and thyme
2 tablespoons finely chopped parsley
½ teaspoon finely grated lemon rind
¼ pint stock (chicken or meat)
1 egg

Remove the rinds from the bacon. Roughly chop bacon, turkey liver and heart and mince them through the coarse blades of a mincing machine. Peel and very finely chop the onion. Melt the butter in a saucepan, add the onion and cook over a low heat until the onion is soft and transparent.

Combine the sausagemeat, bacon, heart, liver, onion and melted butter in a bowl with the breadcrumbs and mix with the hands until the ingredients are evenly distributed.

Beat the egg and mix it into the ingredients with the stock, lemon rind and herbs and season with salt and pepper, beating with a wooden spoon until the mixture is smooth and the liquids have been absorbed.

To freeze: Shape into a brick, free freeze until solid. Pack in a polythene bag, seal, label and freeze.

To use: Thaw stuffing for 12 hours in a refrigerator before using. (Any leftovers can be formed into small balls and fried to serve with the bird as an accompaniment or with cold turkey.)

Storage time: Maximum one month.

Chestnut stuffing for turkey

1 lb chestnuts
2 oz fresh white breadcrumbs
1 oz butter
2 eggs
¼ teaspoon dry mustard
salt and freshly ground black pepper
1 small onion
stock

Cut a cross in the top of the chestnuts. Bake the chestnuts in a hot oven (450°F, 230°C, Reg. 8) for about 15 minutes until the skins are brittle enough to remove. Remove the shells and the inner brown skin of the chestnuts when they are cool enough to handle (do not leave them to get cold as the skins will then become more difficult to remove).

Place the peeled chestnuts in a saucepan and cover them with stock. Bring to the boil and simmer until the chestnuts are soft and most of the liquid has been absorbed. Drain off any excess liquid and mash the nuts.

Peel and finely chop the onion. Melt the butter in a small saucepan. Add the onion and cook over a low heat until the onion is soft and transparent. Add the onion and melted butter to the chestnuts, mix well and leave to cool.

Add the breadcrumbs and mustard to the chestnut mixture and season with salt and pepper. Beat the eggs, add them to the stuffing mixture and mix really well until all the ingredients are fully incorporated.

To freeze: Shape the mixture into a brick and free freeze until solid. Pack in a polythene bag, seal, label and freeze.

To use: Thaw in the refrigerator for 12 hours until soft. Stuff in to the neck end of the bird (any leftovers can be formed into rolls and fried to serve as an accompaniment).

Storage time: Maximum of 4 weeks.

Savoury stuffing balls

(*Serves 4*)

These take little time to make and not only help to add interest to a dish but also help to stretch the main ingredients. The balls can be made and fried and then frozen. They can be re-heated straight from the frozen state.

6 tablespoons parsley and thyme stuffing
6 tablespoons sage and onion stuffing
1 large beaten egg
salt and freshly ground black pepper
½ teaspoon finely grated lemon rind
dripping or lard for frying

Mix the stuffings together, add the egg and enough lukewarm water to make a firm mixture. Add the lemon rind, season with salt and pepper and mix well. Form the stuffing into small balls and chill them in a refrigerator until they are firm.

Heat $\frac{1}{4}$ inch of dripping or lard in a heavy frying-pan. Add the stuffing balls and cook over a medium high heat, shaking the pan, until the balls are crisp and golden brown on all sides.

To freeze: Drain on kitchen paper and leave to cool. Free freeze and, when solid, pack into polythene bags. Seal, label and freeze.

To use: Place the frozen balls on a lightly greased baking sheet and re-heat in a hot oven (400°F, 200°C, Reg. 6) for about 20 minutes until crisp and hot through.

Puddings

Fresh, stewed and puréed fruit all freeze well making an almost instant basis for hundreds of different pudding variations. Pies, tarts, crumbles and mousses can also be stored in your freezer for use at a later date. It is well worth doubling the quantities of almost any pudding you make and freezing half to use when time is at a premium.

My own treacle tart *(Serves 6)*

Each time I write a recipe for treacle tart I seem to come up with yet another improvement. This is the newest yet and I have to admit that it is the best variation yet of a very well tried theme.

I make two or three of these tarts at a time because they not only freeze well but they are also enjoyed by everyone who has ever tasted them. I serve them at the simplest of all summer picnic meals, for plain family lunches or for the most sophisticated of my dinner parties; the reaction is always the same – they go down well and there is never any left at the end of the meal.

8 oz shortcrust pastry (see page 202)
5 tablespoons golden syrup
1 oz butter
2 tablespoons double cream
grated rind of 1 lemon
1 Cox's apple
2 tablespoons Museli or Crunchy Breakfast Cereal

Roll out the pastry to about $\frac{1}{8}$ inch thickness and line an 8/9 inch flan case with the pastry. Brush the bottom of the lined case with a little melted butter to prevent it going soggy during the cooking time.

Combine the syrup and butter in a saucepan and heat gently until the syrup has melted.

Peel and coarsely grate the apple.

Add the apple, cream, lemon rind and Museli or Crunchy Breakfast Cereal to the syrup mixture, mix lightly and spread the mixture in the flan case.

Bake the tart in a hot oven (400°F, 200°C, Reg. 6) for 10 minutes then lower the heat to moderate (350°F, 180°C, Reg. 4) and continue to cook for a further 20 minutes or until the filling has set and the pastry is golden brown.

If the bottom of the flan case rises during the cooking time, prick it lightly with a skewer half way through the cooking time.

To freeze: Cool the tart, free freeze and pack in a polythene bag. Seal, label and freeze.

To use: Thaw at room temperature for 3 hours or until thawed through. Serve the tart warm or cold with cream.

Berry cheese cake *(Serves 6)*

Ring the changes by making the cheese cake with raspberries, black currants, red currants or cranberries.

6 oz digestive biscuits
3 oz unsalted butter
8 oz berries or currants
$\frac{1}{4}$ pint red wine
1 tablespoon gelatine powder
4 oz caster sugar
6 oz cream cheese
finely grated rind of $\frac{1}{2}$ lemon
1 tablespoon lemon juice
1 egg white

Crush the biscuits with a rolling pin. Soften the butter and mix it with the biscuit crumbs. Press the biscuit butter mixture into the base and sides of an 8 inch pie dish and chill in the refrigerator while making the fillings.

Combine the fruit with the wine and simmer until tender. Rub the fruit through a fine sieve and return to a clean pan. Add the gelatine powder and sugar and stir over a low heat until the gelatine and sugar are dissolved. Remove from the heat and leave to cool until the jelly is beginning to set.

Spread the setting jelly over the biscuit crumb base and chill until set firm.

Beat the cheese with the lemon rind and juice until smooth.

Beat the egg white until stiff and fold the egg white into the cheese.

Spread the cheese mixture over the jelly.

To freeze: Free freeze until solid and then pack in a rigid container. Overpack if necessary with a polythene bag, seal, label and freeze.

To use: Thaw in a refrigerator overnight or at room temperature for about three hours.

Lemon and strawberry shortcake* *(Serves 6)*

Step 1

8 oz self raising flour
1 teaspoon baking powder
small pinch salt
3 oz unsalted butter
$1\frac{1}{2}$ oz caster sugar
finely grated rind of 1 lemon
1 teaspoon lemon juice
1 egg, lightly beaten
3 tablespoons milk
$\frac{1}{2}$ oz melted butter

Sift flour, baking powder and salt into a mixing bowl. Rub in the butter until the mixture resembles fine breadcrumbs. Stir in the sugar and half the lemon rind. Add the egg and lemon juice to the dry ingredients with enough milk to make a firm dough. Turn onto a floured board and knead lightly.

Roll out mixture and cut into eight four inch circles using a fluted pastry cutter. Brush four circles with melted butter and top with remaining circles. Bake in a hot oven (425°F, 220°C, Reg. 7) for 12–15 minutes until well risen and golden. Cool for five minutes and then separate the two halves.

Step 2

To freeze: Place the shortcakes on a shallow tray and free freeze until solid. Pack in a polythene bag or rigid container. Seal, label and freeze.

Step 3

To use:

¼ pint double cream	8 oz strawberries
finely grated rind of ½ lemon	icing sugar
1 tablespoon caster sugar	

Thaw the shortcakes in their wrapping for about 2 hours at room temperature (they can be thawed more quickly in a moderate oven for about 20 minutes – leave to cool).

Whip the cream until thick. Halve the strawberries. Lightly fold the lemon rind, caster sugar and strawberries into the cream. Spread the cream over half the shortcakes, top with remaining cakes and dust the tops with icing sugar before serving.

Note: You can substitute raspberries for the strawberries or complete the shortcakes and freeze them with the fillings for a maximum of 1 month.

Orange and lemon mousse (*Serves 6*)

You can use this as a pattern for other fruit mousses. It is easy to make and always popular. Make a quantity of fruit mousses, store them in your freezer and you always have something on hand to serve up as a delicious pudding when the children or guests arrive with friends unexpectedly and you want to produce something special in a hurry.

3 eggs
2 egg yolks
6 oz caster sugar
½ oz gelatine powder
  (1 tablespoon)
¼ pint double cream
juice of ½ lemon
3 tablespoons water
grated rind of 1 orange
1 carton frozen orange juice
2 oz plain chocolate

Separate the whole eggs and combine the yolks and sugar in a bowl. Whisk vigorously with the finely grated orange rind until the mixture is light in texture and pale yellow in colour and the sugar has been dissolved.

Soften the gelatine in the lemon juice mixed with the water and heat over a low flame until the gelatine has completely dissolved. Leave to cool.

Add the orange juice and gelatine mixture to the egg yolk base and mix well. Leave to chill until the mixture is just beginning to set.

Whisk the egg whites until stiff. Whip the cream until light and fairly stiff.

Fold the cream and egg whites into the orange mousse base, turn into a soufflé dish and chill until set firm. Coarsely grate the chocolate and sprinkle it over the top of the pudding.

To freeze: Free freeze the mousse until solid. Overwrap in a polythene bag, seal, label and freeze.

To use: Thaw the mousse in a refrigerator overnight or at room temperature for about three hours.

Serve well chilled with extra cream or a fruit salad.

Lemon chiffon pie

(*Serves* 6)

1 8-inch baked flan case
  (see page 216)
2 eggs – separated
2 teaspoons gelatine powder
2 tablespoons hot water
grated rind of 1 lemon
juice of 2 lemons
6 oz caster sugar

Combine the egg yolks, sugar and lemon juice in the top of a double boiler and whisk over hot, not boiling water until the mixture is light and foaming. Remove from the heat and continue to whisk until the mixture is cool.

Soften the gelatine in the hot water and stir until the gelatine has completely dissolved. Add the gelatine mixture to the egg yolks with the grated lemon rind and mix well.

Whisk the egg whites until stiff. Fold the whites into the lemon mixture and spread in the flan case.

To freeze: Free freeze until solid. Pack in a polythene bag. Seal, label and freeze.

To use: Thaw unwrapped at room temperature for about three hours.

Lemon cheese tartletts (*Serves 6*)

12 oz shortcrust pastry (see page 202)
4 oz ground almonds
2 oz melted butter
½ pint double cream
1 small egg white, beaten
3 egg yolks
1½ oz caster sugar
juice and grated rind of 1 lemon

Roll out the pastry to ⅛ inch thickness and line shallow patty cases. Lightly prick the bottom of the cases and brush them with a little egg white to prevent the cases becoming soggy.

Beat the egg yolks until smooth. Add the almonds, sugar, lemon rind and juice and the melted butter and mix in the cream. Spoon the mixture into the patty cases (do not fill them more than ⅔ full) and bake in a moderate oven (375°F, 190°C, Reg. 5) until the filling has set and the pastry is crisp and golden brown. Leave until cold.

To freeze: Free freeze until solid. Pack in a polythene bag. Seal, label and freeze.

To use: Thaw unwrapped at room temperature for about 2 hours.

Crumbles (*Serves 6*)

Crumbles freeze well and make good use of fruit that is a little past its prime or which is cheap because there is a glut. Use apples, rhubarb, gooseberries, currants, raspberries, apricots etc. (Stew the fruit to remove the pips if necessary before adding the crumble.)

2 lb fruit
6 oz granulated sugar
6 oz plain flour
3 oz butter

Chop fruit like rhubarb and apples, sprinkle over 3 oz sugar and add 2 tablespoons water. Stew fruit that needs to be sieved with 3 oz sugar and 2 tablespoons water.

Sift the flour, add the remaining sugar and mix well. Add the butter, cut into small pieces, and rub the fat into the flour with the fingertips until the mixture resembles coarse breadcrumbs.

Sprinkle the crumble mixture over the fruit pressing it down well to make an even topping. Bake the crumble in a moderately hot oven (375°F, 190°C, Reg. 5) for about 40 minutes until the fruit is cooked and the topping is golden brown.

To freeze: Cool the crumble, pack in a polythene bag, seal, label and freeze.

To use: Thaw the crumble overnight and serve cold with whipped or clotted cream or re-heat in a moderately hot oven (375°F, 190°C, Reg. 5) from the frozen state for about 45 minutes until the fruit crumble is hot through and the topping is nicely crisped.

Note: The crumble can be frozen without being cooked. See also comments on crumble topping on page 202.

Chocolate mousse

(*Serves 6*)

This is a classic chocolate mousse. It freezes well but should only really follow a fairly light main course. It is extravagant but servings should be small.

12 oz plain block chocolate	2 oz caster sugar
1 oz unsalted butter	4 large eggs

Break up the chocolate into a basin, set the basin over a saucepan of hot, not boiling, water and cook over a low heat, stirring occasionally until the chocolate has melted. Add the sugar and stir until the sugar has melted. Remove the chocolate from the heat and beat in the butter. Beat the egg yolks, beat them into the chocolate mixture, and continue beating until the mixture is quite smooth and leave to cool.

Whip the egg whites until stiff and fold them lightly into the chocolate mixture. Turn into individual pots.

To freeze: Cover the pots with foil, overwrap in a polythene bag, seal, label and freeze.

To use: Thaw the mousses in the wrapping for about 1 hour at room temperature or for three hours in a refrigerator. Spoon over a little whipped cream and serve well chilled.

Variations: Add 1 tablespoon Cointreau, Grand Marnier, rum or brandy to the chocolate mixture, or the finely grated rind of $\frac{1}{2}$ orange and 1 tablespoon concentrated frozen orange juice.

Spiced apple pie

(*Serves* 6)

12 oz shortcrust pastry
$1\frac{1}{2}$ lb cooking apples
3 oz sugar
pinch ground cloves
cinnamon and mixed spice
$1\frac{1}{2}$ oz unsalted butter

Cut off a third of the pastry for the top. Roll out the remaining pastry to $\frac{1}{8}$ inch thickness and use it to line a 2 pint capacity pie dish or foil dish.

Peel, core and slice the apples. Combine the sugar with the cloves, cinnamon and mixed spice and mix well. Add the apples and toss them in this mixture so that the slices are coated with the sugar and spice. Cut the butter into small pieces.

Arrange the apple slices in the pie dish and dot the layers with the small pieces of butter. Sprinkle over any remaining sugar mixture.

Roll out the remaining pastry. Cut strips of pastry to go around the sides of the dish. Press the strips onto the top of the pastry in the dish and then cover with the pastry topping. Brush the edges with water, press firmly together, trim off excess pastry and flute the edges with the back of a fork.

To freeze: Cover pastry topping with foil and free freeze. Pack in polythene bag when solid, seal, label and return to freezer.

To use: Remove wrappings and place frozen tart in a hot oven (400°F, 200°C, Reg. 6) for 20 minutes. Cut steam vents in the top of the pastry, brush with beaten egg and continue to cook for a further 30 minutes or until pastry topping is golden brown and pie is cooked through (the exact timing will depend on the depth of the pie).

Alternative fillings: Substitute rhubarb for the apples and add a little more sugar. Pour some cream through the air vents into the top of the pie before serving.

Use half apples and half blackberries when these are in season, or half apple and half tinned apricots.

Flavour the apple with some preserved ginger in syrup, cut into very small pieces, and use some of the ginger syrup instead of the sugar.

Apple purée*

(*Serves* 6)

A basic well flavoured purée to freeze when cooking apples are cheap.

Step 1

$1\frac{1}{2}$ lb cooking apples or windfalls
$\frac{1}{2}$ pint sweet cider
3 oz sugar
pinch cinnamon and ground cloves
1 tablespoon cornflour
1 strip lemon peel

Peel, core and roughly chop the apples. Mix the cornflour to a smooth paste with one tablespoon of the cider.

Put the apples in a saucepan with the sugar, ground cinnamon and cloves, the cider and the lemon peel. Bring to the boil, cover and simmer gently for about 30 minutes until the apples are very soft. Remove the lemon rind and purée the apples through a fine food mill or a fine sieve or in an electric liquidiser.

Return the purée to a clean pan. Add the cornflour paste and cook, stirring all the time over a low heat until the mixture has thickened and is smooth.

Step 2

To freeze: Cool the apple purée, pack it in a rigid container, seal, label and freeze.

Step 3

To use: Turn the frozen purée into a saucepan and heat through over a low heat, stirring, until the purée has thawed through. Or thaw the purée in its container overnight.

Use the purée in any of the following ways:

1. Topped with whipped cream and scattered with roasted almonds. Serve well chilled.
2. Spread in a baked pie shell, top with meringue and bake until the meringue is firm and crisp.

3. With some sliced apricots as a filling for an apple and apricot flan or a pie with a double crust.
4. Add the juice of a lemon and 2 whipped egg whites to the apple purée, turn into a tray or shallow dish, half freeze, beat until smooth and then re-freeze until solid. Serve as a sorbet.

Pears steeped in cider* (*Serves 4*)

Step 1

Pears cooked in red wine have long been a popular pudding dish for the freezer. This is a less expensive and equally delicious version of the same dish.

4 large or 8 small cooking pears
4 oz granulated sugar
4 cloves
pinch cinnamon
½ pint sweet cider
8 oz raspberries

Peel, quarter and core the pears. Place the pears in a saucepan, add the sugar, cloves, cinnamon and cider, bring to the boil and simmer gently for about 40 minutes until the pears are transparent and tender.

Remove the pears, add the raspberries to the sauce, bring the sauce back to the boil and boil over a high heat for 10 minutes. Rub the sauce through a fine sieve or fine food mill to remove the raspberry pips.

Step 2

To freeze: Cool the pears and the sauce. Place the pears in a rigid container and pour over the sauce. Seal, label and freeze.

Step 3

To use:

¼ pint double cream
2 oz flaked almonds

Thaw the pears in their container overnight or for at least 5 hours. Roast the flaked almonds in a hot oven for 1 minute or until golden. Cool the almonds. Whip the cream until thick. Divide the pears between four glass goblets, pour over the sauce. Top with whipped cream and sprinkle over the roast almonds.

Serve well chilled.

Spiced apple cake

You can serve this cold as a cake or heat it through and serve warm with cream as a pudding.

8 oz plain flour
2 teaspoons baking powder
2 oz butter
2 oz caster sugar
4 tablespoons milk
few drops vanilla essence
1 lb cooking apples
4 tablespoons apricot jam
pinch cinnamon

Sift the flour and baking powder into a bowl and rub in the butter with the fingertips until the mixture resembles fresh breadcrumbs. Add the sugar and vanilla essence and mix to a firm dough with the milk.

Press the dough into a well greased Swiss-roll tin raising the edges a little around the sides.

Peel and thinly slice the apples and arrange them on top of the dough in overlapping layers. Heat the apricot jam in a saucepan until it has melted. Mix in a pinch of cinnamon and spread the jam evenly over the apples.

Bake the cake in a hot oven (400°F, 200°C, Reg. 6) for 30 minutes. Leave until cold.

To freeze: Free freeze until solid, slide into a polythene bag, seal, label and freeze.

To use: Unpack and leave to thaw at room temperature for about two hours.

To re-heat: Put partially thawed cake into a moderate oven (350°F, 180°C, Reg. 4) for 15–20 minutes until warmed through.

Mincemeat and apple flan (*Serves 6*)

8 oz shortcrust pastry
2 tablespoons caster sugar
2 cooking apples
12 oz mincemeat
2 tablespoons brandy

Peel and slice the apples. Mix the mincemeat with the brandy. Roll out the pastry to ⅛ inch thickness and line a 8–9 inch flan case. Sprinkle the sugar over the base of the case.

Arrange the sliced apples on the base and spread over the mincemeat. Make a lattice of any remaining pastry to go over the top, brush with beaten egg, and bake in a moderately hot oven (400°F, 200°C, Reg. 6) for about 30 minutes or until the pastry is golden brown and crisp and the apples are soft.

To freeze: Remove pie from flan case. Cool, cover top of pie with foil and 'free freeze' until solid. Pack solid pie in a polythene bag, seal, label and return to freezer.

To use: Remove wrappings and place frozen pie in a moderately hot oven (375°F, 190°C, Reg. 5) for about 30 minutes or until hot through.

Note: Protect the top of the pie with foil if it gets too brown during the reheating period.

The pie can be left to thaw in its wrappings at room temperature and then eaten cold.

Ice Cream

When you have a deep freezer a whole new world is opened up in the field of ice cream making. No longer is there any necessity to empty the ice cube trays from the top of your refrigerator when you want to make ice cream, all you do is pop the mixture into your freezer. Experiment with different flavours, textures and colours and try your hand at making exotic 'bombes' and ice cream shapes.

With a freezer in your home ice creams of all kinds are the obvious freezing candidates. Experiment with yoghurt ice cream, all varieties of fruit and even with savoury ice cream such as cheese or tomato flavouring. You can, of course, also store commercially frozen ice cream in bulk and the containers make good packaging for your home made mixtures.

Remove the ice cream from the freezer for about fifteen minutes before serving it to allow it to soften slightly and for the flavour to develop.

Although some books recommend that you can freeze ice cream for as long as six months, I myself consider it is safer to freeze home made ice cream for a maximum of two months, especially if you plan to open the receptacle it is stored in and remove portions of the ice cream during that time.

The following recipe is for a standard ice cream base.

Basic Ice Cream (*Makes 1½ pints ice cream*)

½ pint Jersey milk
2 eggs, separated
4 oz caster sugar
few drops vanilla essence
pinch salt
¾ pint double cream

Beat the egg yolks with the milk and sugar and the pinch of salt until the mixture is smooth but not frothy. Transfer the mixture to the top of a double boiler and make a custard by stirring over hot but not boiling water until the mixture thickens enough to coat the back of a wooden spoon. (This will probably take about 20 minutes.)

Remove from the heat and stir until cool.

Flavour the custard with a little vanilla essence and add any other flavouring ingredients at this point.

Beat the cream until thick and fold it into the custard. Beat the egg whites until stiff and fold them into the custard and cream mixture. Turn into trays and freeze until about 1 inch of ice crystals have formed around the edge of the trays. Turn into a bowl, break up with an electric or rotary whisk and beat until the mixture is smooth. Pack in rigid containers, seal, label and freeze.

Coffee ice cream

Add two tablespoons very strong black coffee to the custard.

Praline

Melt 2 oz of the sugar used in the recipe over a low heat and continue to heat without stirring until the sugar has turned golden brown. Add 4 oz roasted almonds, mix well and turn onto an oiled slab of marble or baking tray. Leave to cool and then crush finely with a rolling pin.

Fold the praline into the cooled custard.

Ginger ice cream

Replace 2 oz sugar with 2 tablespoons syrup from preserved ginger. Chop 2–3 pieces preserved ginger and add them to the cooled custard with the syrup.

Banana ice cream

Mash 4 ripe bananas with 1 teaspoon lemon juice until smooth. Add the mashed bananas to the cooled custard.

Brown bread ice cream

Bake 4 oz fresh brown bread in a low oven until crisp. Crush the bread into crumbs with a rolling pin.

Fold crumbs into the cooled custard.

Rum and raisin ice cream

Soak 4 oz raisins in two tablespoons rum for 1 hour and then fold the soaked raisins into the custard base.

Rich chocolate ice cream

In this method the sugar is boiled to a syrup with the water and that mixture is then poured over the egg yolks, beating all the time. It makes a very rich and creamy ice cream; the only trouble is, unless you have an electric mixer or food processor you need someone to help you make the ice cream since, as I have discovered to my cost, it is impossible to both pour the syrup onto the egg yolks and beat the yolks at the same time.

7 oz Bourneville chocolate	3 egg yolks
2 oz granulated sugar	1 tablespoon brandy
4 fl oz water	1 pint single cream.

Beat the egg yolks until they are smooth. Dissolve the sugar in the water in a saucepan, bring to the boil, stirring every now and then, and cook over a high heat until the sugar will form a thread when dropped from a wooden spoon (about 20 minutes).

Whisking the egg yolks vigorously, pour the syrup in a slow, steady stream onto the egg yolks. Continue to whisk until the mixture is light and airy and the mousse base has cooled.

Break up the chocolate into small pieces, put it into a small saucepan and add the cream. Cook over a low heat, stirring now and then, until the cream is just below boiling point and the chocolate has melted. Remove from the heat, stir well and leave to cool.

Add the chocolate mixture to the mousse base with the brandy and chill before freezing.

To freeze: Spoon the mixture into a suitable, rigid, container. Seal, label and freeze.

Variations: Add some very strong coffee to the mixture with the brandy to make a delicious coffee/chocolate flavour.

Add some concentrated orange juice and some orange rind to the ice cream with the brandy or use rum or Grand Marnier instead of brandy.

Fruit ice cream

Use fresh or frozen fruit.

Red or blackcurrant ice cream

6 oz blackcurrants	4 egg yolks
1 tablespoon water	$\frac{1}{4}$ pint cream
4–5 oz caster sugar	

Combine the fruit and water and simmer for 5 minutes until the fruit is tender. Press the fruit through a sieve or a fine food mill to remove the pips. Leave to cool.

Combine the yolks and sugar in a basin or the top of a double boiler. Place over hot, not boiling, water and whisk over the heat until the mixture is light yellow and is light and fluffy in texture (this takes time so do not be impatient – when the mixture is ready you should be able to leave a trail through it with the whisk). Remove the bowl from the heat and leave for three minutes.

Whip the cream until thick. Lightly fold the fruit purée and cream into the yolk mixture.

To freeze: Turn into a rigid container, seal, label and freeze.

To use: Remove from the freezer five minutes before required and scoop out with an ice cream scoop or a large spoon.

Variations: You can use raspberries or blackberries in the place of the currants but add a little lemon juice to the blackberries to bring out their flavour.

Orange and lemon sorbet

(*Serves 4–6*)

2 teaspoons gelatine powder
½ pint water
8 oz caster sugar
½ pint fresh orange juice
juice of 2 lemons
2 egg whites

Soak the gelatine in two tablespoons of water to soften. Combine the remaining water and sugar in a saucepan. Bring to the boil and boil over a very high heat for 5 minutes. Remove from the heat, add the softened gelatine mixture and stir until the gelatine has all dissolved. Leave to cool.

Add the orange and lemon juice to the syrup, pour into trays and freeze until ice crystals have formed to a depth of about 1 inch around the sides of the tray. Turn the sorbet into a bowl, break it up and then beat with a rotary or electric whisk until smooth. Whisk the egg whites until stiff and fold them into the sorbet mixture.

To freeze: Turn into rigid containers. Seal, label and freeze.

To use: Turn out of the container and leave to stand for about 10 minutes in the refrigerator before serving.

Variations: Use the equivalent quantity of red currant or black currant juice or thin raspberry or strawberry purée in the place of the orange juice.

Fill half orange shells or lemon skins with the mixture before freezing and remove from the freezer 20 minutes before serving to allow the sorbet to soften slightly. Keep in a refrigerator during this time.

Maidenwell ice cream cake

(*Serves 6*)

This is one of my favourite recipes. The pudding is very versatile and although it is heavy on the double cream it goes a long way.

Use home made meringues or those relatively inexpensive meringue baskets you can buy. I often make the cake when I have leftover egg whites to use up.

1 pint cream (use double cream)
8 oz meringues (broken ones are fine)
3 oz caster sugar
1 tablespoon Kirsch

Whip the cream until thick. Break up the meringues into small pieces. Flavour the cream with the Kirsch and fold in the meringues and sugar.

To freeze: Pack the mixture into an oiled cake tin with a removable bottom. Cover with foil, seal in a polythene bag, label and freeze.

To use: Press the cake out of the tin about 10 minutes before serving and keep it in the refrigerator until you are ready to serve the cake. Cut into wedges and serve with a fresh fruit salad, raspberries, strawberries or stewed fruit.

Variations: Add some chopped, preserved, ginger and some ginger syrup to the mixture at the same time as you fold in the meringues.

Sprinkle some chopped, roasted, almonds over the bottom and sides of the cake tin before packing in the mixture.

Mix some pralines, or some peeled grapes with their pips removed, into the cream with the meringues.

Make the cake in a ring mould and fill the centre with fresh fruit or a colourful ice cream or sorbet.

Pancakes

Pancakes are amongst the best of all freezer material and having a good stack of them in your freezer means you have access to any number of very quick and delicious meals.

The basic pancake mixture can be used for either sweet or savoury pancakes, they can be rolled and served with a little grated cheese or some caster sugar over them and they can be piled up in a stack with a filling between each layer. A good rich creamy or cheese sauce can be poured over the pancakes or you can serve the savoury variety with a tomato sauce and the sweet pancakes with a custard sauce or with fresh dairy cream.

Pancakes should be as thin as you can make them – the thinner they are the better they taste. Do not be upset if your first pancake doesn't turn out to be a winner. The first one often sticks and tends to be too thick. Providing you are using a good pan (non stick pans are ideal for this), the second pancake should come easily away from the pan and if the mixture is too thick you can always thin it with a little milk or water.

You can freeze the pancakes with or without fillings. I find it best to freeze them without fillings as it is really the making of the pancakes, not the fillings, which takes the time and frozen pancakes without fillings are very swiftly re-heated from the frozen state.

Pile the pancakes onto a plate when they are cooked, putting a strip of greaseproof paper between each one. Wrap them in polythene bags (I usually pack the pancakes in batches of six or eight at a time), seal, label and freeze.

To use: Wrap the frozen pancakes loosely in foil and re-heat them in a moderate oven (350°F, 180°C, Reg. 4) for about 20 minutes or until they are thawed through.

Smaller pancakes can be re-heated in a frying pan with a little butter or bacon fat and served with maple syrup and small sausages as a delicious Sunday breakfast or 'brunch' dish.

The basic pancake recipe*

(Serves 4 (about 8 pancakes))

Step 1

4 oz plain or self raising flour	1 large egg, lightly beaten
pinch salt	½ pint milk

Sift the flour into a bowl and add the salt. Using a wooden spoon make a hollow in the centre of the flour and drop in the lightly beaten egg. Gradually work the egg into the flour, adding the milk a little at a time, until the mixture is fairly smooth. Finish with a rotary whisk, beating until the lumps have all been worked out and the mixture is bubbly and the consistency of thin cream.

To make pancakes in an electric liquidiser or food processor
Combine the ingredients in an electric liquidiser or food processor and process until the mixture is smooth. You may find that you can add a little extra milk to pancakes made in this way to thin the mixture to the required consistency.

The thinner the pancake mixture the more delicious and crisp the pancakes will be.

To cook
Pancake batter; lard, olive oil or a good vegetable oil for frying.

Use a non-stick omelette pan and do not attempt to cook the pancakes in a pan which you use for frying other ingredients.

Put about a teaspoon of lard or oil into the pancake pan and heat over a light heat until the fat begins to smoke. Add one tablespoon of batter and swirl it around the pan so that the bottom of the pan is evenly coated with a thin film of batter (use a jug for pouring the batter). Cook for 1 minute over a high heat, until the bottom of the pancake is golden brown. Turn the pancake over with a spatula and cook on the other side for 1 minute until the second side is golden. Slide the pancakes onto a plate, stacking them one on top of another as they are cooked.

Step 2
To freeze the pancakes unfilled: Put a sheet of greaseproof paper, Clearfilm or foil between each pancake and wrap the whole stack in Clearfilm or foil. Pack them in a polythene bag, seal, label and freeze.

Do not store for longer than 3 months.

Step 3
To use: Wrap the frozen pancakes loosely in foil and re-heat them in a moderate oven (350°F, 180°C, Reg. 4) for about 20 minutes or until they are thawed through, or thaw for about 3 hours at room temperature. Use any of the following recipes.

1. Savoury pancakes

Chicken or turkey pancakes *(Serves 4)*

Do not store for longer than 1 month. 12 pancakes, thawed at room temperature for three hours or, wrapped in foil, in a moderate oven (350°F, 180°C, Reg. 4) for about 20 minutes until soft.

1 onion
4 oz firm button mushrooms
1½ oz butter
1½ tablespoons flour
½ pint milk
10–12 oz cooked chicken or turkey
2 tablespoons finely chopped parsley
salt and freshly ground black pepper
4 tablespoons double cream
2 oz grated Gruyère cheese

Peel and finely chop the onion. Slice the mushrooms very finely. Melt the butter in a saucepan, add the onion and cook over a low heat until the onion is soft and transparent. Add the mushrooms and cook for three minutes. Blend in the flour and gradually add the milk, stirring continuously over a medium high heat until the sauce comes to the boil and is thick and smooth.

Chop the chicken or turkey. Fold the chicken or turkey and the parsley into the sauce and season with salt and freshly ground black pepper. Blend in the cream and remove at once from the stove.

Spread the filling on the pancakes, roll them up neatly, cover with the grated cheese and bake in a hot oven (400°F, 200°C, Reg. 6) for 15 minutes until the pancakes are hot through and the topping is golden and bubbling.

The pancakes can be made and then frozen with the filling

Fill the pancakes and place them in a baking dish or foil container. Cover with foil and pack in a polythene bag. Seal, label and freeze. Do not store for longer than 1 month.

To cook the frozen stuffed pancakes

Unpack the frozen pancakes, sprinkle with cheese, cover with foil and bake in a moderately hot oven (350°F, 180°C, Reg. 4) for about 40 minutes until the pancakes are hot through – remove the foil for the last 10 minutes of cooking time so that the top becomes golden and bubbling.

Crab stuffed pancakes (*Serves 4*)

8 frozen pancakes, thawed at room temperature for 3 hours or in a low oven (300°F, 150°C, Reg. 2) until soft.

1 small onion
1 small green pepper
2 oz butter
12 oz mixed white and brown crabmeat
2 oz white breadcrumbs
2 teaspoons finely chopped parsley
salt, freshly ground black pepper and a pinch cayenne
few drops Tabasco sauce
few drops Worcestershire sauce
2 teaspoons dry English mustard
¼ pint double cream
2 oz grated Gruyère or Cheddar cheese

Peel and finely chop the onion. Remove the core and seeds of the pepper and finely chop the flesh. Heat the butter in a saucepan, add the onion and pepper and cook over a low heat until the onion is soft and transparent. Add the crabmeat, breadcrumbs and parsley, season with salt and pepper, mustard and cayenne and flavour with a few drops of Tabasco and Worcestershire sauce. Cook for 2 minutes and stir in the cream.

Fill the pancakes with the crab mixture and roll them up neatly. Place the rolled pancakes in a lightly greased baking dish, sprinkle over the cheese and bake in a moderately hot oven (375°F, 190°C, Reg. 5) for about 20 minutes or until the pancakes are hot through and the topping is golden brown.

The pancakes can be made with the filling and then frozen
Fill the pancakes with the filling and place them in a greased baking dish or foil container. Cover the pancakes with foil. Pack them in a polythene bag, seal, label and freeze.

Cook the frozen stuffed pancakes as recipe for Chicken Stuffed Pancakes.

Haddock with cheese sauce (*Serves 4*)

8 pancakes thawed (7 in)
1 lb smoked haddock
2 hard boiled eggs
1½ oz plain flour
¼ pint cream
3 oz grated Cheddar cheese
½ pint milk
2 bay leaves
2½ oz butter
salt and white pepper
pinch of mace

Place the haddock in a shallow pan with the bay leaves and pour over the milk. Bring slowly to the boil and simmer for 10 minutes or until the haddock is just tender. Strain off the milk, remove any skin and bones from the haddock and flake the flesh. Chop the hard boiled eggs; add 1 oz butter, the cream and the eggs to the haddock and mix lightly. Divide the haddock mixture between the pancakes and roll them up lightly. Place the pancakes in a lightly greased baking dish.

Melt the remaining butter in a saucepan, add the flour and mix well. Gradually blend in the milk, stirring continually until the mixture comes to the boil and the sauce is thick and smooth. Add 2 oz grated cheese and stir over a low heat until the cheese has melted. Season with pepper and a little mace and pour the sauce over the pancakes.

Sprinkle over the remaining cheese and heat through under a grill or in a moderate oven until the cheese on top has melted and is golden brown and the pancakes are hot through.

Pancakes with Bolognese filling (*Serves 4–6*)

8 thawed pancakes (7 in)
1 portion thawed sauce Bolognese (pages 134–5)
1½ oz butter
1½ oz plain flour
½ pint milk
3 oz Cheddar cheese
1 oz Parmesan cheese
salt and white pepper
pinch of nutmeg

Fill the pancakes with the thawed and heated Bolognaise mixture. Roll them up and place the pancakes in a lightly greased baking dish.

Melt the butter in a saucepan. Add the flour and mix well. Gradually add the milk, stirring over a moderate heat until the sauce comes to the boil and is thick and smooth. Add the Parmesan and half the Cheddar cheese, season with salt, pepper and a pinch of nutmeg and stir over a low heat until the cheese has melted. Pour the sauce over the pancakes, top with the remaining cheese and heat in a moderately hot oven (400°F, 200°C, Reg. 6) until the pancakes are hot through and the cheese topping is golden and bubbling.

Fisherman's filling (*Serves 4*)

4 thawed pancakes (7 in)
4 oz peeled prawns
2 spring onions
1½ oz butter

1½ oz flour
8 oz white fish (cod, coley, haddock)
½ pint milk
salt
freshly ground pepper
2 oz grated Cheddar cheese

Place the fish in a shallow pan and cover it with the milk, season with salt and pepper, bring to the boil and simmer gently for about 10 minutes until the fish is just tender. Strain off the milk, remove the skin if necessary from the fish and flake the fish.

Very finely chop the spring onions.

Melt the butter, add the flour and mix well. Gradually add the milk, stirring continually over a medium heat until the sauce comes to the boil and is thick and smooth. Add the prawns, fish and onions and season with salt and pepper.

Fill the pancakes with the fish mixture and roll them up neatly. Place the pancakes in a shallow lightly buttered dish, sprinkle over the cheese and grill under a grill until the cheese has melted and the pancakes are hot through.

Other fillings for savoury pancakes

Remember to cover the stuffed pancakes with a white or cheese sauce or to top the filled pancakes with grated cheese to prevent them drying out too much during the cooking time.

Combine two packets frozen spinach, cooked until tender and well drained, with ricotta cheese, season with salt, pepper and a pinch of grated nutmeg and cover the pancakes with a cheese sauce or with a generous sprinkling of grated cheese.

Fold prawns into a rich white sauce flavoured with some Vermouth and finely chopped nuts.

Thinly slice some button mushrooms, cook them in some butter and fold them into a white sauce with some crisply fried, crumbled bacon and top with grated cheese.

Use a basic meat sauce (see pages 134–5) and top with grated Parmesan or Gruyère cheese.

For a special occasion, fill one end of the pancakes with cream cheese and the other end with red caviar and top with grated Gruyère cheese.

2. Sweet pancakes

Lemon pancakes (*Serves 4*)

8 pancakes
juice of 2 lemons
2 tablespoons caster sugar

Sprinkle the lemon juice over the pancakes and roll them up neatly. Place the pancakes in a shallow, lightly buttered baking dish and heat through in a hot oven (400°F, 200°C, Reg. 6) for about 10 minutes or until the pancakes are piping hot. Sprinkle over the sugar and serve at once.

Pancakes jubilee (*Serves 4–6*)

This is a very exotic pudding but it must be made at the last minute and served at once. The pancakes can, of course, be made in advance, frozen and thawed.

8 pancakes
1 block vanilla ice cream
8 tablespoons black cherry jam
4 tablespoons caster sugar

Spread the pancakes with the black cherry jam. Cut the ice cream into eight slices. Place a slice on each pancake, roll them up quickly. Place them in a shallow baking dish, sprinkle over the sugar and brown under a very high grill until the sugar has melted and is bubbling and turning brown.

Crêpes suzette (*Serves 4*)

Make your pancakes on the small size and as thin as possible.

8 small, thin, pancakes
3 oz unsalted butter
grated rind of 1 orange
grated rind of 1 lemon
juice of 2 oranges
juice of 1 lemon
4 oz caster sugar
1 tablespoon rum or brandy

Beat the butter with the sugar until creamy and pale in colour. Gradually blend in the orange and lemon peel and the fruit juices and brandy (do not worry if the mixture curdles).

Spread the filling over the pancakes, fold in half and then in half again to make triangles and arrange the pancakes in a shallow, fireproof serving dish.

To freeze: Cover the dish tightly with foil and overwrap in a polythene bag. Seal, label and freeze.

To use: Take the dish out of the polythene bag and put it into a hot oven (400°F, 200°C, Reg. 6) and heat through for about 30 minutes until the sauce is bubbling and the pancakes are piping hot. Heat 2 tablespoons brandy in a large spoon over a candle, set the brandy alight, pour it over the pancakes and serve them at once.

Other fillings for sweet pancakes

Fill the pancakes with tinned black cherries, drained and stoned, or with black cherry jam. Heat through and serve with cream.

Combine cream cheese with some sugar and finely-chopped nuts and raisins and fill the pancakes with this mixture. Sprinkle over some caster sugar before serving.

Fill the pancakes with apricot jam, or a mixture of apple and apricot purée. Heat through and sprinkle with lemon juice and caster sugar.

Fill the pancakes with stewed and sweetened red and black currants and heat through. Sprinkle generously with caster sugar before serving and accompany with double or clotted cream.

Purée some stewed apple, combine the purée with some finely-grated lemon rind, some sugar to sweeten and some finely-chopped raisins. Or add chopped, roasted almonds or some finely-chopped preserved ginger. Fill the pancakes with the apple mixture, heat through and sprinkle with caster sugar before serving.

Pastry, Bread, Biscuits and Cakes

Once you decide to do some baking it is really very little extra trouble to double the quantity of the mixture you are making so that half can be frozen. Raw or cooked pastry, cooked or uncooked bread mixes, biscuits and cakes all freeze well providing a valuable store of products to have at your fingertips.

Almost all pastry freezes well, either cooked or raw. The exception to this rule is the hot-water crust that is usually used for raised pies. This is inclined to absorb ingredients and become flabby in the freezer.

Shortcrust, flaky and puff pastry all freeze well uncooked but since puff pastry is such a chore to make I would recommend buying one of the first class commercial varieties on sale rather than attempting to make your own. You can also buy good commercially frozen flaky pastry.

Choux pastry is better if it is cooked before freezing and the éclairs or profiteroles can then be filled after thawing. Eclairs or profiteroles can also be filled before freezing but they should not be stored in the freezer for longer than a month.

Basic shortcrust pastry

This is a bulk quantity of pastry. Make it when you have time to spare in the kitchen and can make a quantity of pies to freeze for a later date.

1½ lb plain white flour
1 teaspoon salt
6 oz lard
6 oz butter or margarine
iced water to mix

Mix the flour and salt together in a large bowl. Add the lard and butter cut into small pieces and rub the fats into the flour with the fingertips. When the mixture resembles fairly coarse breadcrumbs, add just enough iced water to form the mixture into a firm, smooth dough with your hands. Roll out on a well floured board.

Note: 8 oz of the mixture can be frozen as a crumble topping for quick fruit puddings. Add 2 oz sugar, a pinch of cinnamon and mixed spice to the flour and fat mixture and do not add any water. Freeze the crumble mixture in polythene bags and sprinkle it, straight from the freezer, over a fruit mixture.

Bake the crumble in the normal way.

To freeze: Divide the pastry into suitable portions and shape each portion into a rectangle. Pack in polythene bags, seal, label and freeze.

To use: Thaw at room temperature for about 2–3 hours.

To freeze lined flan cases before cooking

It is useful to keep some lined flan cases in the freezer, ready to be filled, if you have space enough.

Line the flan case in the usual way, free freeze until solid and then overwrap in polythene bags.

To use: Fill the frozen case with your chosen filling and cook in the normal way.

To freeze cases that have been 'baked blind' before filling

These are not so hardy as uncooked flan cases and care should be taken to ensure that the pastry will not get damaged in the freezer.

Bake the cases 'blind' (roll out the pastry thinly and line your flan case. Line the case with greaseproof paper or foil and weight it down with dried peas or beans. Bake the pie in a hot oven (400°F, 200°C, Reg. 6) for 10 minutes, remove the greaseproof paper or foil lining and continue to bake in a moderate oven (350°F, 180°C, Reg. 4) until crisp and pale golden brown (about 10 minutes). Cool, free freeze and then pack in a polythene bag (I use cardboard plates to protect the cases).

To use: Remove the cases from their packaging materials, fill and bake in the normal way. If the filling is to be one that does not need cooking, thaw the case for two hours at room temperature before adding the filling.

Note: Cases that become a little soft through freezing can be crisped-up in a moderately hot oven for a few minutes.

To freeze dishes with fillings – uncooked

A number of good pastry dishes, both sweet and savoury, can be made up to the cooking point and then frozen, e.g. apple pie, steak and kidney pie, mincemeat and apple pie. Fill the pastry with a cooled filling, do not cut a vent in the pastry topping, free freeze, pack, seal, label and freeze.

To use: Unpack and bake in the normal way allowing an extra 45 minutes for thawing, cutting a vent in the top and brushing with beaten egg once the pie has thawed.

To freeze pastry dishes with fillings – cooked

Cook pastry dishes in the normal way. Cool, free freeze, pack in polythene bags, seal, label and freeze.

To use: Thaw at room temperature for 4 hours and re-heat in a moderate oven to serve hot.

Note: Good foil cases can now be bought for baking pastry dishes and these are ideal if you wish to store your cases or pastry dishes for any length of time.

Quiches

These French savoury tarts make good starters, light lunch or supper dishes or they can be served cold for picnics and outdoor meals. They benefit from being re-heated before serving.

Basic recipe (*Makes 2 × 7 to 8 inch flans*)

12 oz short crust pastry (see page 202)
2 tablespoons vegetable oil
2 large onions
4 oz streaky bacon rashers
4 oz firm button mushrooms
3 eggs
¼ pint single cream
¼ pint milk
2 oz Gruyère cheese, grated
1 oz Parmesan cheese, finely grated
salt and freshly ground black pepper
1 tablespoon finely chopped parsley

Roll out the pastry thinly and line two 7–8 inch flan cases. Peel and very thinly slice the onions. Remove the rinds from the bacon and chop the rashers. Slice the mushrooms.

Heat the oil in a heavy frying pan. Add the bacon and cook over a low heat until the fat runs out. Add the onions to the bacon and continue to cook over a low heat until the onion is soft and transparent. Remove the bacon and onions with a slotted spoon and leave to cool. Add the mushrooms to the juices in the pan and cook over a medium high heat for 2 minutes, stirring. Leave the mushrooms to cool.

Arrange the onions, bacon and mushrooms over the bottom of the flan cases. Beat the eggs until smooth. Add the milk and cream and season with salt and pepper and mix in the parsley and cheese.

Pour the egg mixture over the ingredients in the flan case and bake in a hot oven (400°F, 200°C, Reg. 6) for 15 minutes. Reduce the heat and continue to bake in a moderate oven (350°F, 180°C, Reg. 4) for a further 30 minutes or until the filling is firm and the pastry is golden brown. Leave to cool.

To freeze: Free freeze the flans until solid. Pack them in polythene bags, seal, label and freeze.

To use: Unwrap, place the frozen flan on a baking sheet and re-heat in a moderate oven (350°F, 180°C, Reg. 4) for about 30 minutes until hot through. The flans can be left in their wrappings and thawed overnight.

Variations: Parboil some leeks and drain them well. Combine leeks and bacon in the flan case and pour over the egg mixture as above to make a traditional *Quiche Lorraine.*

Cook some small cocktail sausages and arrange them in the flan case with some fried onions and a little bacon. Pour over the egg mixture as above.

Combine some cooked spinach with some cottage cheese, season well, add a little cooked bacon and pour over the egg mixture as above.

Add some chopped ham to the basic ingredients.

Add some cooked flaked haddock to the basic ingredients and omit the bacon.

Add some white crabmeat to the basic ingredients and omit the bacon.

Profiteroles and savoury puffs*

Another basic commodity like pancakes that are invaluable to keep in the freezer to use as delicious, quickly made dishes that can be both sweet or savoury and which can be prepared with very little notice.

The basic recipe for choux pastry is the same whether it is to be used for sweet or savoury dishes. The puffs can be frozen with or without their fillings and can be served hot or cold depending on their fillings.

Basic choux pastry recipe*

Step 1

2½ oz plain flour
2 oz butter
pinch salt (or use salted butter)
¼ pint milk and water mixed
2 eggs well beaten

Combine the water, milk and butter in a saucepan, bring to the boil and simmer until the butter has all melted. Add the flour, all at once, and beat hard, off the heat, until the mixture forms a smooth paste and comes away from the sides of the pan. Leave to cool for 5 minutes and then gradually beat in the beaten eggs, a little at a time until the mixture is smooth and glossy.

Spoon the mixture onto a greased baking sheet – 1 tablespoon for large puffs, 1 teaspoon for small puffs – or pipe through a ½ inch plain nozzle into 3 inch lengths, cutting off the required length with a sharp knife. Bake in a moderately hot oven (400°F, 200°C, Reg. 6) for 20–40 minutes until the puffs are well risen, golden brown and sound hollow when tapped. It helps to turn off the heat when the puffs are cooked and leave them in the oven with the door just ajar to dry out for 10–15 minutes.

Remove the puffs from the baking sheet, cut a slit through their sides and pull out any soft dough from the middle.

Step 2

To freeze: Place the puffs on a shallow tray and freeze them until they are solid. Pack the puffs in a polythene bag, seal, label and freeze.

Step 3

To use: Thaw the frozen puffs, in their bag, for about 2 hours at room temperature or place the frozen puffs on a baking sheet and thaw them in a moderate oven (350°F, 180°C, Reg. 4) for about 20 minutes.

Cool the puffs if necessary, fill them and serve hot or cold.

Note: The puffs can be filled and frozen with their filling. In this case re-heat the frozen puffs for about 40 minutes in a moderate oven (350°F, 180°C, Reg. 4) or, if the puffs are to be served cold, leave them to thaw at room temperature for about 4 hours.

I find it better to freeze puffs without fillings as they do have a tendency to go soggy if they are to be served cold and re-heating them for a short time crisps them up very quickly.

Éclairs

(Makes 12/16 éclairs)

Step 3 ingredients

12/16 cooked frozen piped choux pastry puffs piped through a ½ inch plain nozzle in three inch lengths
½ pint double cream
1 tablespoon caster sugar
4 oz icing sugar
1/2 tablespoons warm water
2 teaspoons cocoa powder

Thaw the pastry puffs at room temperature for about 2 hours or in a moderate oven (350°F, 180°C, Reg. 4) for about 20 minutes until crisp and thawed and leave to cool.

Sift icing sugar into a bowl. Add the cocoa powder to the warm water and mix until smooth. Add the cocoa mixture to the icing sugar, a little at a time, mixing well until the icing is smooth.

Whip the cream until thick and lightly fold in the caster sugar (the cream can be very lightly flavoured with vanilla essence).

Fill the éclairs with the whipped cream and top them with the icing brushed on with a pastry brush.

Note: The éclairs can be filled and then frozen, with or without icing. To thaw the filled éclairs, leave them unwrapped at room temperature for about an hour; they are best eaten while still slightly frozen.

Cherry filled profiteroles with caramel sauce

(Serves 4)

12 small choux pastry puffs thawed
8 oz black cherry jam
4 oz granulated sugar
4 tablespoons water

Thaw the puffs at room temperature and crisp them in a moderate oven (350°F, 180°C, Reg. 4) for about five minutes and leave to cool (or thaw in a moderate oven until crisp and thawed through, for about 15 minutes) and leave to cool.

Fill the puffs with black cherry jam.

Put the sugar into a small heavy saucepan and cook over a low heat without stirring until the sugar has melted and is golden brown. Add the water (stand back because this process is inclined to spit), and mix until a thick caramel has formed. Leave to cool for 15 minutes.

Pile the profiteroles on a serving dish, pour over the caramel and serve at once.

Note: This dish is also delicious if the profiteroles are filled with a mixture of black cherry jam and ice cream or just ice cream only. Fill the profiteroles at the last minute (or fill them with ice cream and then return them to the freezer for a maximum of 1½ hours before serving) pour over the caramel and serve at once.

You can also fill the profiteroles with whipped cream and pour over a hot chocolate sauce at the last minute. Melt 4 oz plain slab chocolate in a basin over hot water, stir in 1 small tin of evaporated milk or single cream and pour the hot chocolate sauce over the profiteroles.

Bread

Basic white bread (*Makes 2 large or 4 small loaves*)

3 lb strong white flour
1 oz salt
1 oz lard
4 teaspoons dried yeast
1 teaspoon sugar
1½ pints warm water

Mix the flour and salt together in a large bowl. Rub in the fat, cut into small pieces, until it has virtually disappeared.

Combine ¼ pint of the water, warmed to blood heat, with the sugar and mix until the sugar has melted. Sprinkle the dried yeast over the surface and leave in a warm place until the mixture is spongy – about 10–15 minutes.

Add the yeast mixture and the remaining water to the flour mixture and mix well with the hands.

Turn the bread mixture onto a floured board and knead with the knuckles until smooth and elastic. Put the dough in an oiled bowl, cover with a polythene bag (also oiled), leaving plenty of room for the dough to rise and double in bulk.

Leave the dough in a warm, draught free place until it has doubled in bulk (about 40 minutes).

Turn dough onto a floured board and knead lightly for 5 minutes. Shape dough into two oblongs, tuck them into oiled baking tins, seam side down and put inside a greased polythene bag (close the bag but allow room for the dough to rise) for a further 30 minutes in a warm place. Brush the loaves with water and bake in a hot oven (425°F, 220°C, Reg. 7) for 20 minutes. Turn loaves round and reduce the oven temperature to 400°F, 200°C, Reg. 6 and continue to bake for a further 20 minutes or until loaves that are turned out sound hollow when tapped on the bottom. Cool the loaves on a wire rack.

To freeze: Pack the cooled loaves in polythene bags, seal, label and freeze.

To use: Thaw at room temperature for 2 hours. For a crisp crust bake the thawed bread for about 10 minutes in a hot oven (400°F, 200°C, Reg. 6).

Note: To make toast the frozen loaf can be cut into slices with a special freezer knife and the slices can be toasted in the normal way.

Variations: ½ quantity of white bread mix; 2 large onions; 2 tablespoons vegetable oil.

Follow the instructions for making 'basic white bread'. Peel and very finely chop the onions. Heat the oil, add the onions and cook over a medium low heat until the onions are crisp and golden brown. Drain the onions on kitchen paper and add them to the made bread mixture before leaving it to rise for the first time.

Cottage loaf

Use ⅓ of the basic bread dough. Cut off a third to make the top of the loaf and shape the larger piece into a ball. Flatten the ball slightly and top with the remaining dough also shaped into a ball. Keep the top in place by brushing the bottom of it with a little milk.

Flour the handle of a wooden spoon and push it through the centre of both balls. Pull the handle out sharply, cover the loaf loosely with a well oiled polythene bag and leave in a warm place until the loaf is well risen. Dust with flour (or brush with milk if you like a crisp crust) and bake in a hot oven (425°F, 220°C, Reg. 7) for 25 minutes.

To freeze: Cool on a wire rack. Pack in a polythene bag, seal, label and freeze.

To use: Thaw at room temperature for two hours.

Baps

The perfect containers for hamburgers especially when you are cooking the burgers over a barbecue.

1 lb strong white flour	pinch sugar
1 teaspoon salt	$\frac{1}{4}$ pint milk
2 oz butter or margarine	$\frac{1}{4}$ pint water
$\frac{1}{2}$ oz fresh yeast	

Cream the yeast with a pinch of sugar. Warm the milk and water together until they are at blood temperature. Add the warm liquid, a little at a time, to the yeast, stirring all the time so that the mixture is smooth and making sure that you stir in all the yeast. Leave in a warm place while preparing the dough.

Sieve the flour into a bowl with the salt. Add the butter cut into small pieces and rub the butter into the flour with the finger-tips until the mixture resembles fine breadcrumbs. Mix in the yeast mixture until the dough becomes stiff and fairly smooth. Turn the dough onto a well floured board and knead for about 10 minutes until smooth and shiny.

Place the dough in a large well greased polythene bag, seal lightly leaving plenty of room for expansion and leave the dough to rise in a warm place for about one hour or until doubled in bulk.

Turn out onto a floured board and knead again until the dough is firm. Divide into 10 pieces, shape into balls and then flatten each ball in half using the palm of the hand. Sprinkle the top of the baps with sifted flour and leave to rise in a warm place for about 20 minutes, until doubled in bulk again.

Bake in a very hot oven (450°F, 230°C, Reg. 8) for 15 minutes until firm and lightly golden on top. Cool on a wire rack.

To freeze: Free freeze the baps until solid and then pack in polythene bags, seal, label and freeze.

To use: Thaw at room temperature for 1 hour or heat through in a moderate oven (350°F, 180°C, Reg. 4) for about 15 minutes until thawed through.

Wholemeal bread

(Makes three loaves)

3 lb brown wholemeal flour
1 oz salt
1 oz lard
4 teaspoons dried yeast
1 teaspoon sugar
$1\frac{1}{2}$ pints warm water

Mix the flour and salt in a large bowl and rub in the fat with the fingertips. Stir the sugar into $\frac{1}{4}$ pint warm water (it should be at blood heat temperature) and sprinkle the dried yeast onto the top of the liquid. Leave in a warm place until frothy and bubbling – about 10 minutes.

Add the yeast liquid and remaining water to the flour and mix well. Turn onto a floured board and knead with the knuckles until the dough is smooth and elastic. Place dough in an oiled bowl, cover with an oiled polythene bag (allow plenty of room for the bread to rise) and leave in a warm place until the dough has doubled in bulk – about 1 hour.

Turn onto a floured board and knead to knock out all the air. Shape the dough into three oblong loaves, put into well greased bread tins and cover with oiled polythene. Leave to rise until doubled in bulk and then bake in a moderately hot oven (400°F, 200°C, Reg. 6) for about 1 hour or until the loaves sound hollow when tapped on the bottom.

To freeze: Cool the loaves on a wire rack. Pack the cooled loaves in polythene bags, seal, label and freeze.

To use: Thaw at room temperature for two hours. For a crisp crust bake the thawed bread for about 10 minutes in a hot oven. (400°F, 200°C, Reg. 6).

Variations

Poppy seed rolls

Divide a third of the dough into ten pieces and shape each one into a ball. Flatten the balls slightly, cover lightly with oiled polythene and place the rolls on an oiled baking sheet. Leave to rise for about 30 minutes until doubled in bulk. Brush the top of the rolls with milk, sprinkle over some poppy seeds and bake in a moderately hot oven (400°F, 200°C, Reg. 6) for about 30 minutes.

To freeze: Cool on a wire rack. Free freeze until solid and then pack in polythene bags, seal, label and freeze.

To use: Thaw at room temperature for one hour.

Tea ring

Something a little different to serve with tea.

⅓ basic white dough mix	3 oz icing sugar
4 oz mixed dried fruit	2 tablespoons lemon juice
1 oz demerara sugar	1 oz flaked almonds
pinch mixed spice	1 oz glacé cherries

Leave the dough to rise for the first time (see recipe on page 208). Roll out the dough to an oblong about 14 in × 10 in and sprinkle over the dried fruit, spice and demerara sugar to within 1 inch of the edge of the dough. Damp the edges, roll up neatly and form into a circle sealing the ends with water. Put the circle on a baking sheet and into an oiled polythene bag and leave to rise in a warm place until doubled in bulk.

Remove from the bag and bake in a hot oven (400°F, 200°C, Reg. 6) for about 30 minutes or until golden brown and sounding hollow when tapped.

Cool on a wire rack and cover with an icing made by mixing the icing sugar with the lemon juice.

Toast the flaked almonds in a hot oven for 1 minute until golden brown. Cut the glacé cherries into slices. Press the almonds and glacé cherries into the top of the loaf.

To freeze: Free freeze the loaf until solid and pack in a polythene bag. Seal, label and freeze.

To use: Remove from the wrapping and thaw at room temperature for about two hours before cutting.

Note: You can heat the thawed loaf in a moderate oven (350°F, 180°C, Reg. 4) for about 15 minutes until warm before serving.

Cheese plait

⅓ mixture of basic bread dough	4 oz grated strong Cheddar cheese
1 teaspoon dry mustard	celery salt

Add the mustard and cheese to the bread dough when kneading it and before leaving it to double in bulk for the second time.

Divide the dough into three pieces and roll each piece into a long sausage shape. Plait the pieces together, brush ends with beaten egg and secure them firmly together. Place on a greased tray and put inside a loosely tied, oiled polythene bag. Leave to rise until doubled in size. Remove from the bag and sprinkle with celery salt and bake in a hot oven (400°F, 200°C, Reg. 6) for about 30 minutes until golden brown and sounding hollow when tapped on the bottom.

To freeze: Leave to cool on a wire rack. Pack in polythene bag, seal, label and freeze.

To use: Put the frozen loaf into a moderately hot oven (375°F, 190°C, Reg. 5) and bake for about 20 minutes until hot through and crisp on the outside. Or leave to thaw at room temperature for about 2 hours.

Croustades

(Makes 6 croustades)

A quickly made alternative to puff pastry *vol-au-vents.* Made a number at a time and store them in the deep freeze. Fill with any of the suggestions for filling vol-au-vents or savoury pancakes.

6 medium slices white bread
oil

Remove the crusts from the bread and flatten them slightly with a rolling pin to prevent the bread breaking up. Grease 6 muffin tins and press the bread gently into the tins. Brush the surface of the bread with oil and bake them in a hot oven (400°F, 200°C, Reg. 6) until they are crisp and golden. Leave to cool.

To freeze: Pack in rigid containers, seal, label and freeze.

To use: Fill the frozen croustades with a cold, cooked filling and heat through in a moderate oven (350°F, 180°C, Reg. 4) for 20–30 minutes until the croustades and filling are hot through.

Pizzas

(Makes 2 pizzas)

⅓ basic white dough mix
1 tablespoon vegetable oil
1 large onion
1 tin tomatoes (15 oz)
1 tablespoon tomato purée
salt and freshly ground black pepper
pinch oregano and basil
4 thin slices Gruyère cheese

Peel and very finely chop the onion. Heat the oil in a saucepan, add the onion and cook over a low heat until the onion is soft and transparent. Add the tomatoes and tomato purée, season with salt and pepper and mix in the herbs. Bring to the boil, mix well, breaking up the tomatoes with a fork, and cook over a low heat for 15 minutes (do not cover – the liquid should almost disappear by the time you come to use the filling). Leave to cool.

After the dough has risen the first time divide it into two pieces and roll out each piece into a circle about 9 inches in diameter. Pinch up the edges a little to make a ridge around the outside.

Spread the filling over the pizzas, top with the Gruyère cheese and free freeze until solid.

To freeze: Wrap in Clearfilm and overwrap in a polythene bag. Seal, label and freeze.

To use: Unwrap the pizzas, put them onto a lightly oiled baking sheet and bake in a hot oven (400°F, 200°C, Reg. 6) for about 30 minutes until the pizzas are hot through and the cheese has melted and is golden brown.

Alternative fillings for the pizzas

Cover the top of the pizzas with thin slices of salami and some thinly sliced, stuffed, olives before adding the cheese.

Place a layer of thinly sliced, lightly fried mushrooms before adding the cheese.

Place some anchovies, cut into halves, lengthwise, on top of the cheese.

Cover the tomato purée with some drained sardines and some slices of Mozzarella cheese.

Cover the tomato purée with thin slices of ham before adding the cheese.

Roll the dough into two oblongs and sandwich together with some softened, sliced, onions (fry them in butter until soft) and some anchovies. Brush the top with milk and cool and freeze as above.

Freezing home made cakes and biscuits

Nearly all cakes freeze well providing you follow some simple guidelines. Next time you make your favourite cake recipe, double the quantities so that you can eat one today and save the other, in your freezer, for a later date.

Guidelines to cake freezing

1. Loaf-shaped cakes are easier to store than circular cakes and taste just the same.
2. It is better not to ice the cake before freezing it.
3. Thaw cakes in their wrappings.
4. If cakes are to be frozen with their icing on they will need good protection in the freezer so pack them in rigid containers, filling up any air spaces around the cake with crumpled kitchen paper.
5. Always make double the quantity of any cake you plan to make. It will not take you any more time to increase the proportions and you can eat one when it is baked and keep the other in the freezer.
6. A good many biscuit mixes freeze well and many can be cut from the frozen state into thin slices and baked at once making extra specially crisp and light biscuits.

Brack or tea bread

A rich tea bread which freezes extremely well.

½ lb sultanas
¼ lb raisins
¼ lb stoned dates, chopped
1 lb soft brown sugar
1 pint tea
½ pint dry cider
1 lb plain flour
3 teaspoons baking powder
3 eggs, beaten
2 teaspoons mixed spice

Combine the tea (it should be good and strong) with the cider, pour the liquid over the sultanas, raisins and dates and leave to steep overnight.

Combine the flour and baking powder in a large mixing bowl. Add the sugar and mix well. Add the mixed spice, beaten eggs and the soaked fruit and stir until the ingredients are well blended. Turn into a well greased 2 lb loaf tin (or two 1 lb tins) and bake in a moderately low oven (325°F, 170°C, Reg. 3) for 1½ hours or until a knitting needle plunged into the centre of the bread comes out clean. Turn out and cool on a wire rack.

To freeze: Cool, pack the bread in polythene bags, seal, label and freeze.

To use: Thaw at room temperature for about 2 hours.

Sponge flan cases

(Makes 1–8 inch flan case)

3 oz plain flour
3 eggs
3 oz caster sugar
1 oz melted butter

Whisk the eggs and sugar together in a bowl over hot but not boiling water until the mixture is thick and the whisk leaves a trail over the surface when lifted from the mixture. Remove from the heat and continue to whisk until the mixture is cold. Fold the flour lightly into the egg mixture with the melted butter. Turn into a well greased and floured 8-inch flan ring and bake in a moderately hot oven (375°F, 190°C, Reg. 5) for 20 minutes or until firm. Turn out onto a wire rack and leave to cool.

To freeze: Free freeze the flan case until solid, pack in a polythene bag, seal, label and freeze.

To use: Thaw at room temperature for about 2 hours and then fill with a cooked mixture.

Saturday cake

A delicious and easy way to make cake with a crisp topping.

8 oz self-raising flour
1 teaspoon baking powder
3 oz caster sugar
juice and grated rind of 1 lemon
4 oz glacé cherries
1 egg
4 oz butter
scant $\frac{1}{4}$ pint sweet cider

The topping:
2 oz soft brown sugar
1 oz butter
fine browned breadcrumbs
1 oz self-raising flour

Oil an 8 inch cake tin with a removable bottom and coat the sides and bottom with a thin layer of dried breadcrumbs (these will stick to the surfaces that have been oiled).

Sift the flour and baking powder into a bowl. Add the butter, cut into small pieces, and rub the fat into the flour with the fingertips until the mixture resembles coarse breadcrumbs. Mix in the sugar, lemon rind and the glacé cherries, halved. Lightly beat the egg with the lemon juice and enough cider to make ¼ pint liquid until the mixture is smooth. Add the liquid to the cake ingredients and mix until blended. Turn the mixture into the cake tin.

Mix 1 oz flour with the brown sugar. Melt the butter and mix it into the flour and brown sugar. Spread this mixture lightly over the top of the cake and bake in a moderately hot oven (375°F, 190°C, Reg. 5) until the cake has risen and is golden brown on the top. Leave to cool for five minutes before turning out onto a wire rack.

To freeze: Leave to cool, pack in polythene bag, seal, label and freeze.

To use: Leave to thaw, still in its wrapping for two hours until thawed through.

A Freezing Calendar

A month by month list of items to look out for with your freezer in mind and ideas for pre-cooking food in season for deep freezing.

January and February

Seville oranges are in season. If you do not have time to make marmalade when the oranges are on sale pack the oranges in polythene bags, seal, label and freeze. See recipe on pages 94–5.

Winter vegetables can be made into soups or vegetable purées or can be chopped and blanched to use later in stews and casseroles.

Freeze small, firm Brussels sprouts from your garden. Brussels sprouts also make a good puréed vegetable.

Mandarins, clementines and satsumas etc. can be peeled, divided into segments with skin and membrane removed and frozen in a medium sugar syrup.

Tangerine juice makes an excellent flavour for water ices.

March

Fish is often readily available, reasonably priced and in good condition this month. Find out where your nearest fishmarket is and buy the fish in bulk, cutting out the middlemen or ask the advice of your fishmonger and see if you can come to some arrangement with him to clean and perhaps fillet the fish. Make fish stock from the bones and trimmings of the fish to use later as the liquid in which you cook your fish or as a base for the sauces that are to go with it.

Early rhubarb should be around and can be frozen raw, with or without sugar, or it can be stewed ready to serve. Rhubarb makes a delicious flavouring for ice cream.

April

Fish should still be plentiful and in good condition and there should also be plenty of broccoli around at this time of the year. Prime broccoli is so delicious it can be served with an *hollandaise* sauce as a first course or use it instead of spinach as a base for a Florentine dish with eggs and a cheese sauce.

Fennel is often around too at this time of the year and the puréed bulbous root makes an excellent flavouring for soups or to serve as a vegetable.

May

Spring clean your freezer now before you want to start stocking it with the new season's produce.

After the first burst of spring, home produced lamb, although not cheap, is particularly delicious and succulent. Spring lambs are not large and it might well be worth your while buying a whole carcass for your freezer. It is worth paying a little extra to get an experienced butcher to cut up the lamb for you.

New carrots and the first small peas should be coming in now and both these vegetables freeze really successfully.

June

Grab those first French beans when they are really small, slender and very tender. Asparagus can sometimes be bought astonishingly cheap and freezes well – good bargains can be had by buying second quality asparagus which makes an excellent base for soup.

Cherries are not the most satisfactory freezers of all, but peaches will be on sale now and they freeze well in a fairly thick sugar syrup. Watch out for small, tasty melons and gooseberries to make a delicious sauce to go with fish.

July

Buy cheap strawberries to make purée to flavour ice creams and to use as a sweet sauce.

Raspberries should be getting into full swing and are a must for the freezer. Free freeze them before packing to preserve their shape. Loganberries are also delicious and a good cropper. Try growing some in your garden – they have a sharper and more full bodied flavour than raspberries.

Fruit is generally most plentiful at this time of the year and so make a point of freezing a good variety to keep you going during the winter months when it is not so plentiful.

August
Aubergines and cheap tomatoes are around now, together with late courgettes or those rather dreary marrows. Take the opportunity to make masses of aromatic ratatouille to serve cold as a first course or with cold meats or to serve hot as a vegetable or as a stuffing for green or red peppers.

Runner beans make a cheap vegetable for any month of the year and they freeze well. You can also purée runner beans and a handful of very finely chopped runner beans are delicious in a mixed vegetable soup. Do not cut your beans too fine for freezing or they will loose some of that essential crispness.

There should be some imported seedless grapes about and these are very useful to freeze for adding to fruit salads or for making that classic dish of '*fish véronique*'. Cover the grapes with a slight sugar syrup.

Blackberries are coming into fruit now. They need no preparation and a day spent in the country combing the hedgerows can pay great dividends later when you have the fruit on hand for making succulent old fashioned apple and blackberry pie.

September
Go on picking blackberries because on the 1st October, so they say, the devil spits on the fruit and it should no longer be picked.

See if you can get hold of a brace of grouse or two. Hang game before freezing and prepare it for the oven. Thaw game completely before cooking it. Old birds can often be bought for quite a reasonable price and make delicious rich casseroles.

There should be plenty of fish around but be careful. If the weather is very hot the condition of the fish you buy in the shops is not likely to be quite as fresh as it really should be for freezing. If you buy fish from a wholesaler make sure you get it home and into the freezer as quickly as possible.

Vegetables and fruit should all be in good condition and not too expensive. Plan out your freezing life carefully so that you neither become a slave to your deep freeze or freeze ingredients which you are going to regret later.

October and November
Plenty of game around now. Try to find a small supplier if your family are fond of game so that you can buy it more cheaply than by buying it from a large store. You will usually be able to save quite a bit by plucking the birds yourself and once you have tried

your hand at this a few times you will find it can really be done in quite a short time.

The first Christmas turkeys may be coming into the shops now and although it may seem a bit early to be thinking about Christmas you can make a great saving by buying your bird early before the prices begin to rise sharply in the last few weeks before the holiday.

Make your Christmas pudding and cake and think about pre-cooking and freezing dishes for Christmas in the last few weeks of November. Christmas always seems to creep up on one and the freezer can be the greatest possible asset in cutting down the work you have to do for the holiday.

Chestnuts can be cooked and puréed to serve as a vegetable, mixed with Brussels sprouts, as a stuffing for the Christmas bird or for use in sweet dishes.

New season's nuts keep better in the freezer than they do drying at room temperature. Pack the nuts in their shells and thaw them at room temperature for three or four hours.

Watch out for fresh cranberries to make a cranberry sauce for Christmas.

If the weather is good watch out for bargains in the fish line and think about buying some meat in bulk to last you through the winter. English lamb can be a very good buy at this time of the year.

December

Cook ahead for those parties you are going to give over Christmas and the New Year. Check through your freezer and throw away any produce that is over six months old and which you are not going to use.

Work out how long it is going to take to thaw your turkey and do not forget to take it out of the freezer in plenty of time.

Glossary of Freezing Terms

Like cookery, the art of deep freezing has developed some terms of its own which may be slightly confusing to the novice. The following list of terms may help to clarify any terms you are not quite sure about.

Accelerated freeze drying: This is a technical term which has no connection with home freezing and is a commercial method of drying foods such as coffee and some herbs.

Air spaces: If air is allowed to circulate around the inside of packs that contain solid ingredients the produce will deteriorate in the freezer due to dehydration; as much air as possible therefore should be excluded from such packs. If the produce is packed in a polythene bag or other soft container the excess air should be pressed out before the bag is sealed. If the produce is stored in a rigid container any air spaces around the ingredients should be filled with crumpled kitchen paper.

Liquid ingredients, on the other hand, expand when frozen so it is necessary to leave room for this expansion by allowing an air space at the top of containers used to store such liquids as stock, soups, sauces etc.

Airtight seal: By expelling all possible air and sealing your freezer packages properly you should be able to attain an airtight seal in your packaging material.

Alarms: If your freezer is situated in a garage or some place where you may not visit it at least once a day it is a sensible precaution to invest in a battery operated alarm which will be set off if the temperature of the freezer rises above the safety point due to a power failure or mechanical fault.

Ascorbic acid: Synthetic vitamin C which can be bought in powder form from chemists and is diluted and added to some foods when preparing them for the freezer to replace loss of vitamin C and prevent the ingredients from discolouring during their processing, for example, delicate fruit that is inclined to brown like peaches and nectarines; hard fruit which is inclined to brown like apples and pears, and vegetables which are inclined to discolour when peeled like celeriac, globe artichokes and Jerusalem artichokes.

Baskets: Most freezers are equipped with baskets when you buy them but, in my experience, you always need to buy one or two more – the more you can separate and categorize the contents of your freezer the easier it will be to find the produce you want in a hurry and the more efficient your use of the freezer will be.

Extra baskets can be bought from your freezer manufacturer or from suppliers of deep freeze packaging materials.

Batch baking: Make the best possible use of your oven when you are baking – don't just cook one loaf or one cake, double or treble the quantities and cook now to use later.

Blanching: By blanching vegetables in boiling water for a few minutes you arrest the action of enzymes in produce, prevent loss of colour and the deterioration of the texture, flavour and nutritional value of the produce. See section on the 'Preparation of vegetables for the deep freeze'.

Blanching basket: A useful piece of equipment if you plan to do a lot of home freezing of garden produce, this is a fairly fine mesh, wire basket into which prepared vegetables are put for blanching and cooling. By using a blanching basket you avoid the necessity of having to re-heat a new lot of water for each batch of vegetables you have to blanch.

Blast freezing: Food frozen on a commercial scale is often 'freeze blasted' by being passed through a tunnel of highly refrigerated air. The advantage of this method of freezing is that the ingredients stay separate during freezing – some manufacturers refer to this method of freezing as 'free flow' on the outside of their packs.

Bulk buying: By owning a freezer you have the opportunity of buying both fresh and frozen food in bulk and thereby saving quite a considerable amount of money. Care must be taken not to buy too much fresh produce at one time so that you are not able to prepare and process it properly for freezing and you should also make sure that the produce you are going to buy in bulk, whether it is fresh or commercially frozen food, is something your family are going to be happy to eat in quite large quantities. Take advantage of seasonal prices and gluts when buying in bulk and work out how much space you actually have in your freezer before making large purchases that you find you cannot store when you get them home.

Catering packs: Large packs of frozen food which are usually sold for catering establishments but which can prove to be an economy for the housewife provided she buys produce her family is going to enjoy, the quality of the produce is to her liking and the pack is divided up into family sized portions before being put into the freezer or as soon as it is opened for the first time.

Block freezing: If you want to package liquid ingredients in polythene bags it is advisable to pour the liquid into a rectangular container, freeze until solid and then turn out and package in your bag making a neat package that is convenient to store – this is referred to as 'block freezing'.

Chest freezers: Usually more bulky than upright freezers but cheaper to run and easier for the storage of large and bulky products like large joints of meat and poultry. Chest freezers are available in sizes of 100–900 litres (3·5–32 cubic feet) and are usually sold with wire dividing panels and baskets.

Chinagraph pencils: Coloured marking pencils which will write on any surface and are not affected by damp. These pencils are ideal for labelling almost any packaging materials used in the freezer and their only disadvantage is that it is almost impossible to wipe off the writing if the packaging material is to be re-used.

Crystalisation: All food contains some liquid and this forms ice crystals in the produce when it is frozen. The faster the produce is frozen the smaller will be the crystals. Large crystals break up the fibres of food affecting both texture and flavour. Fresh produce should be dried carefully before freezing.

Defrosting of the deep freeze: Some modern deep freezers have an automatic de-froster; all other freezers should be defrosted at least once every six months or when a layer of ice has formed on the inside walls of the freezer. If any ingredients are actually spilt in your freezer it is worth defrosting the cabinet and giving it a good clean up as soon as possible to avoid the tainting of other food by any odours which may escape into the freezer. See page 12 for instructions on how to defrost your freezer.

Dehydration: If food is not properly wrapped when it is frozen, moisture from the ingredients will be absorbed into the cold air of the freezer and the produce will quite swiftly lose both texture and flavour. Food affected from this lack of moisture suffers from 'freezer burn'.

Discolouration of frozen food: Some produce will discolour during its freezer life unless proper care has been taken to avoid this. Discolouration occurs in produce which turns brown when it is cut (i.e. some fruit, aubergines, some root vegetables, avocados and cauliflower). Avoid this by sprinkling the cut surface of the produce with lemon juice, by immersing fruit in a sugar syrup and vegetables into a light brine. Asorbic acid, lemon juice or white vinegar can be added to the water in which fruit and vegetables liable to discolour are blanched or cooked.

If food is inadequately packaged it will often discolour during freezing, vegetables will become dull in colour and meat or poultry are inclined to get grey patches on the surface.

Dry ice: Solid carbon monoxide used to keep frozen food at a low temperature during travel. Dry ice can be bought from some retailers of frozen food but it should never be handled with bare hands, come into contact with unwrapped food or be put into the deep freeze.

Dry pack: Packaging food for the freezer without adding liquid or some kind of sugar additive.

Dry sugar pack: Fruit packed with dry sugar rather than in a sugar syrup or dry pack.

Enzymes: Substances which work in all raw foodstuffs causing deterioration and which need to be inactivated by blanching to prevent discolouration and lack of colour and texture.

Expansion: Food that is sealed in an airtight container which it is then to be cooked in should have a vacuum space in the pack to allow for the expansion of air during re-heating. Without this vacuum space the bag or container being used will break.

Liquid expands as it freezes therefore any liquids stored in the freezer should be packed with room for expansion – allow for at least 2·5 cm (one inch) expansion room at the top of narrow containers in which liquid foods are to be stored.

Fan cooled freezer: Some of the larger freezers have a fan incorporated into the main body of the freezer to circulate the cold air evenly throughout the machine. Although this system undoubtedly makes for more efficient freezing, machines with a fan are on the noisy side as the fan rotates all the time even when the motor itself is not running.

Fast freezing: The faster food can be frozen the better will be the results especially in long term freezing. Always use the fast freeze switch when you are going to freeze fresh produce, turning it on 1–2 hours before placing the food in the freezer. Do not freeze too much fresh produce at one time and remember to switch off the fast freeze switch as soon as the produce is frozen solid.

Free flow freezing: (also referred to as free-frozen and open frozen) Another term found on commercially frozen food which refers to items that are blast frozen separately and therefore, in the case of items like peas and cut beans will 'flow freely from their package'.

Free flow freezing can be done in the home and is very much recommended for peas, beans and soft fruit. Special plastic trays for this purpose can be bought from suppliers of deep freeze packaging materials. The prepared ingredients to be frozen are then carefully dried to remove all possible moisture, spread on the trays in a single layer and frozen as quickly as possible without being covered. As soon as the produce is frozen solid it should be packed with all possible speed and returned to the freezer.

Freezer burn: One of the hazards of freezing which should be avoided at home and carefully watched for when buying commercially frozen meat, fish, or poultry. Dehydration of inadequately packed food, allowing air to reach the frozen produce results in oxidation, the produce will dry out, fat tends to become rancid and white patches appear on the surface of the frozen product; meat affected in this way becomes grey and loses its texture and flavour.

Frost free freezers: Some of the more modern freezers are fitted with an automatic defrosting mechanism which operates about once every six hours. Frost free freezers still need to be emptied and cleaned out at least once a year to remove any odour that may have built up inside the freezer but the chore of cleaning out the freezer is obviously much quicker than it would be with a freezer that you have to defrost manually.

Headspace/Headroom: The space left at the top of a container to allow for the expansion of liquid ingredients such as stock, soups and sauces etc.

Insulated frozen food containers: Most large deep freeze food supermarkets now supply insulated bags in which to pack the frozen food you buy and these are ideal for short journeys. If you live a long way from a freezer centre it might be worth investing in an insulated box which will keep the food frozen for a longer period. Frozen food that is to be stored in your own deep freeze should not be allowed to defrost and then re-frozen.

Labelling: This is all-important for successful freezing. You may think you will instantly recognise a package you have just put in the freezer but the chances are that within only a few days you will not be able to tell whether it is a package of chops, some beef sandwiches or a double portion of fish cutlets. Label everything you put in your freezer stating the date it was frozen, the contents and, where applicable, the number of servings the package contains. A wide range of labelling material is available from most stationers or from specialists in deep freeze packaging materials (see page 54).

Microwave cooking: Pre-cooked food can be defrosted and re-heated in an almost miraculously short time in a microwave oven although without a special 'micro-browner' or grill being incorporated into the oven the food will not brown in the same way as it would by more conventional cooking methods.

Moisture – vapour proof packaging materials: Cheap and inadequate packaging materials are amongst the most frequent cause of disappointing results from the freezer. The correct packaging material is specially prepared to be both moisture and vapour proof and therefore protects frozen food from the effects of odours it might pick up while frozen, oxidation and dehydration.

Open freezing: see free flow freezing.

Overwrapped packages: In order to give produce double protection it is often overwrapped in a polythene bag after having been first wrapped in foil, transparent film or other packaging materials. Good airtight packaging is vital if the best results are to be obtained when freezing at home. Any produce with protruding points such as chops, joints with bones, fish with spines etc., should have both an inner and an outer wrapping for double protection.

Packaging: see pages 48–56.

Preformed packs: Liquid foods can be frozen in rectangular shapes which will help you to stack your freezer more efficiently. Line a convenient rigid sugar pack or other container with a polythene bag, pour or spoon in the soup, stew or sauce etc., that is to be frozen, seal, label and freeze. Remove the outer pack when the produce has become solid. Alternatively, soups, stews etc., can be frozen in a rigid polythene container, turned out when they are frozen and then repacked in a polythene bag.

Stews, casseroles and other made up dishes can be frozen in the dish they were cooked in and then turned out and repackaged in polythene bags. When the frozen dish is required, merely turn it back out of the bag into its original container and re-heat in the oven.

Rancidity: When the frozen fat of meat or poultry is exposed to air in the freezer because it has been inadequately packaged the fat, particularly if the meat is salted, will develop an off taste as rancidity sets in and will also suffer from loss of texture and become dry.

Re-heating frozen foods: Almost every form of pre-cooked frozen food requires its own particular method of re-heating to give the best possible results. Some dishes can be put into the oven when they are completely frozen and others should be thawed and then re-heated. Look at individual recipes for detailed instructions

The quickest way to re-heat frozen food is to use a microwave oven but this will not brown the top of the finished dish, unless you use a special browning dish or the oven is fitted with a special browning attachment (see microwave).

Syrup packs: Fresh fruit packed with a sugar syrup; the fruit may be left raw or may be cooked before the syrup is added.

Temperature: The fast freezing temperature of your deep freeze for freezing raw produce should be below 9°F (–23°C).

The storage temperature of frozen food should be below 0°F (–18°C).

Thawing of frozen food: Thawing times and procedures vary for different produce and thawing times depend very much on the temperature the produce is thawed in. Meat, poultry and whole fish should be thawed as slowly as possible, preferably in the bottom of a refrigerator and meat thawed in this way will suffer the minimum of drip or blood loss during the thawing time. Whole poultry must *always* be completely thawed before being cooked, although joints of meat can be cooked from the frozen state providing the temperature of the oven is not over 400°F, 200°C, Reg. 6.

Vacuum packs: Commercially frozen food is often sold as 'vacuum packed' which means that it has been packed under pressure and all air has been excluded from the pack. When buying items such as salt meat, bacon or cured meat for your home freezer it is best to buy such a pack as these are all products which are particularly susceptible to deterioration from oxidization through air being able to get to the surface of the product.

Index